Climate and Society
Climate as Resource, Climate as Risk

Nico Stehr
Zeppelin Universität, Germany

Hans von Storch
GKSS Research Centre, Germany

World Scientific

NEW JERSEY · LONDON · SINGAPORE · BEIJING · SHANGHAI · HONG KONG · TAIPEI · CHENNAI

Published by

World Scientific Publishing Co. Pte. Ltd.
5 Toh Tuck Link, Singapore 596224
USA office: 27 Warren Street, Suite 401-402, Hackensack, NJ 07601
UK office: 57 Shelton Street, Covent Garden, London WC2H 9HE

British Library Cataloguing-in-Publication Data
A catalogue record for this book is available from the British Library.

CLIMATE AND SOCIETY
Climate as Resource, Climate as Risk

Copyright © 2010 by World Scientific Publishing Co. Pte. Ltd.

All rights reserved. This book, or parts thereof, may not be reproduced in any form or by any means, electronic or mechanical, including photocopying, recording or any information storage and retrieval system now known or to be invented, without written permission from the Publisher.

For photocopying of material in this volume, please pay a copying fee through the Copyright Clearance Center, Inc., 222 Rosewood Drive, Danvers, MA 01923, USA. In this case permission to photocopy is not required from the publisher.

ISBN-13 978-981-4280-53-2
ISBN-10 981-4280-53-4

Typeset by Stallion Press
Email: enquiries@stallionpress.com

Printed in Singapore by B & Jo Enterprise Pte Ltd

Acknowledgements

Our book builds on a German version which was translated by Robin Branstator and Hans von Storch, and edited by Paul Malone. We also have to thank Sebastian Belser, a student of Zeppelin University for his assistance in preparing the manuscript. We are thankful to Lona Liesner for processing the manuscript. Last but not least we are indebted to Barbara Stehr for proofreading our text and for the preparation of the index.

The data for Fig. 1 we obtained from Friedrich-Wilhlem Gerstengarbe and Peter Werner; Figs. 2 and 17 was assembled with data drawn from the website of the Danish Meteorological Institute; Figs. 4 and 10 we received from Heiner Schmidt; Fig. 5 from Harold Brooks; Figs. 7 and 8 from Roger Pielke Jr.; Figs. 2, 8, 17 and 25 are assembled or redrawn from information provided on the Internet (see figure captions for their sources); Fig. 12 we obtained from the Geophysical Institute in Bergen; Fig. 16 was also redrawn using information provided on the Internet by the Weather Service of Niger; Fig. 22 is from the Third Assessment Report of the IPCC; the two Christmas-card images in Fig. 23 are reproduced from the article by M. Rebetez; the data in Fig. 24 is from Dennis Bray, and the remaining figures are taken from other books published by Hans von Storch and Nico Stehr. We thank them all. The graphical material has been prepared or enhanced by Beate Gardeike.

Contents

Acknowledgements	v
1. Overview	1
2. A Historical Overview of Thinking about Climate	5
3. Climate as Limiting Condition and Resource	11
3.1. Climate as an Environmental Experience	13
3.2. Climate as a Scientific System	31
3.3. Climate as a Social Construct	43
3.4. Society and Humans as a Climate Construct	46
4. Climate as Risk and Hazard	61
4.1. The History of Ideas about Climate Change	65
4.2. Natural Climate Variability	71
4.3. Excursus: Who Owns the Weather in 2025?	79
4.4. Anthropogenic Climate Change	82
4.5. Climate Change as a Social Construct	95
4.6. The History of Anthropogenic Climate Catastrophes	106
4.7. The Influence of Climate Changes on Society	112
5. Zeppelin Manifesto on Climate Protection	129
6. Summary and Prospects	135
Index	139

1

Overview

The natural climate creates one of the most important general conditions for our existence. For that reason, climate has been one of the most fundamental themes of human reflection for centuries. It has repeatedly been observed that climate not only is the foundation of human civilization, but also causes its particular forms, successes and failures.

Humans are therefore either at a disadvantage or favored, depending upon their climatic region. However, humankind is not merely if at all, and at all times, a creature determined by climate, and climate is not just an object of human contemplation. Climate is also partly the result of human activity, a condition recently increasingly confirmed by the scientific community.

The discussion of anthropogenic (caused by humans) global climate change has become more and more intensive in recent years, and claims that the scientific issues regarding such changes, perspectives and causes are "settled". And it is this connection that almost everyone today understands the concept of the "Greenhouse Effect". An American congressman has declared global warming "the greatest danger for our planet". Public surveys indicate that the public in industrialized countries ranks climate change first among environmental dangers. Scientists appear very alarmed; they address the public directly and warn of an imminent climate catastrophe.

In this book we will discuss these issues and try to embed them into a political, cultural, economic and historical context. Climate is not a novel issue and problem, and while it is broadly defined by scientific concepts, it is also a culturally and socially constructed issue.

In this introductory chapter, we offer a short overview of the important themes of this book, as well as an introduction to the subject "Climate and Society". Climate is, again, a theme discussed in many social institutions. This is true for everyday life and also for science, politics and economics. However, the term "climate" conveys different meanings in these various fields of human activity. On one side there is the scientific concept, believed by many to be the only relevant notion of climate; but there are also various representations of weather phenomena, climate conditions and climatic influences that have arisen over the centuries. The scientific understanding of climate has not annulled or extinguished these widely held and culturally constituted views. Today common sense ideas of climate continue to have an important function in the everyday life of society. Diverse ways of thinking coexist with, and create, social and political actions and reactions. This book tries to sort out and describe the different facets of the concept of "climate".

In the second chapter "A Historical Overview of Thinking about Climate", we describe how this concept has attained a social and political role. We also discuss the extent to which the idea of "climate" affects society and politics; and how it changes, or does not change, throughout history. Early on, people observed a close relationship between climate and society, particularly between climate and human well-being. We will document this relationship. Before the modern "scientification" of the notion of climate, it was rare in previous centuries to speak of climate where people had not, or could not have, settled. One could not, for example, have conceived of a climate of oceans, or of Mars. Early climatology was an auxiliary science of geography, in whose center stood the physiological and psychological effects of climate on people. Today a much more comprehensive concept of climate prevails in science. In the second chapter, we describe this changing understanding of climate over the course of time.

In Chapter 3 ("Climate as Limiting Condition and Resource"), we deal with climate as it manifests itself without human interference. Climate appears as a reliable factor of our environment, providing the conditions for the organization of activities and commerce of individuals and society. It confronts them with calculable risks. The individual can only experience climate as this stationary framework. Climate changes take place on time scales both comparable to and much longer than the human horizon of experience. In this sense, "climate" seems to be like a slot machine that reliably pours out various amounts of money according to fixed rules of probability. One can depend on the fact that seldom (but now and then), large winnings will pour out. Many players expect — irrationally — after a large win (a climate extreme) a long dry streak (climatically unremarkable times). Lengthy observations of winnings from playing (weather) allow for estimating the probabilities (for "normal conditions" and extremes), and rational strategies may be deduced based on expected wins and losses.

In Chapter 4 ("Climate as Risk and Hazard") we no longer consider climate as a "constant" phenomenon, but rather as something variable. Of course in this context the aspect of anthropogenic climate change has recently entered the picture; but we will see that it was almost always there.

In Chapter 5, we bring together the strands of the analysis into our "Zeppelin Manifesto",[1] which spells out in a series of hypotheses what we suggest is needed for the design of a balanced and efficient climate policy. Certainly, the issue of climate and its impact on society is too important to leave it only to natural scientists, who often fail to understand their own conditioning by cultural elements.

---

[1] The term "Zeppelin" refers to the institution *Zeppelin University*, where Nico Stehr holds a professorship.

2

A Historical Overview of Thinking about Climate

The observation and explanation of climatological phenomena can be divided roughly into three important phases. These phases not only fall into historically distinct epochs of different lengths, but also express different interests, methods of observation and explanatory approaches, and also are sometimes aimed at different audiences.

Interest in the climate question manifested itself very early in the history of mankind. In the first phase, humans stood in the center. The original preoccupation with climate always included the search for the effects, as well as the mechanisms, of climate on the human condition, its mood and its health. Climate was a key constraint for human life and civilization.

In the waning years of the 19th century, purely physical ways of thinking about climate began to dominate, at least in science; and that methodological approach made climate research an independent scientific subject. This is the second phase of thinking about climate. This branch of science became relevant for society by providing tables, maps and atlases of climatic averages and the character, type and frequency of extreme conditions required for planning purposes. Climate became "the statistics of weather". In this phase climate was considered neutral, whereas in the first phase climate had been regarded as a more or less efficient existential resource for the people exposed to it.

Today we are experiencing the development of the third phase, in which climate is not only an external given but (within limits) capable of being changed and manipulated by humans. In a certain sense, the contemporary phase harks back to the themes of the first phase. Because the changes are not equally geographically distributed, climate loses its neutrality. There are "winners" and "losers". "Climate Change" becomes a part of the political sphere, in which climatology becomes a key policy tool in achieving societal goals and values.

In recent years, research into the "mechanics" of climate variability has become somewhat less important than research into the climatic effects on ecosystems and social systems. The theme of "climate" has renounced the ivory tower of science, which first describes and then analyses.

Many modern climate researchers no longer perform the role of the basic scientist operating in isolation from society, but are media and policy experts, who impress the populace and political actors with gripping pictures of threatening perspectives about the future living conditions of humans and society.

The German naturalist and explorer Alexander von Humboldt (1769–1859) belonged to the early observers of the first phase. In the first volume of his work *Cosmos, A Sketch of a Physical Description of the Universe*, initially published in 1845, he outlined a physical world description of the concept of climate with the following words: "The term climate, taken in its most general sense, indicates all the changes in the atmosphere, which sensibly affect our organs, as temperature, humidity, variations in the barometrical pressure, the calm state of the air or the action of varying winds, the amount of electric tension, the purity of the atmosphere or its admixture with more or less noxious gaseous exhalations, and, finally, the degree of ordinary transparency and clearness of the sky, which is not only important with respect to the increased radiation from the earth, the organic development of plants, and the ripening of fruits, but also with reference to its influence on the feelings and mental condition of men".

Humboldt's description of the phenomenon of climate not only calls attention to the genesis and state of climate through certain geophysical and atmospheric processes, but also refers to the effects of climate on man's emotional condition and physical well-being.

The revolutionary change in the understanding of climate instituted at the end of the 19th century — and the accompanying growth in the scientific reception of climate — led to a new concept of climate that received more attention than other possible points of reference of climate, underscoring climate's position as "the totality of meteorological phenomena, which characterize the (average) condition of the atmosphere in any position of the earth's surface". (Julius von Hann, 1839–1921)

References to the physical, psychological and social consequences of climate faded, and the quantitative description of climate based on instrumental determinations of climatological variables prevailed. Climate and weather were differentiated: Weather is the transient, real, local atmospheric condition of the day. Climate is the statistics of weather calculated over long periods of time, and usually for larger geographic areas. Or, in modern words, weather is a random process, the properties of which are described by its climate. These statistics are determined from a series of measurements and observations of atmospheric values, primarily temperature, precipitation and wind speed. In particular, the description of average relationships played a decisive role at this time.

The main emphasis of climate research lay in a geographical description comparing different regions, and in the classification of the averages of variable weather conditions over longer periods of time. Climate appeared more or less static. It was geographically limited by the atmospheric boundary layer over land. The global climate was no more than the sum of all regional climates.

Only when scientists were no longer limited to observations at the surface, due to technical innovations in the 1920s, did the third phase of climate research begin. Climatology conclusively became a special branch of science, dealing almost exclusively with the physical description of climatological processes. More and more physicists turned to the investigation of atmospheric and oceanic occurrences. The hitherto traditional link to geography loosened in favor of a new discipline, "Physics of the Atmosphere and/or the Ocean". In the wake of this conceptual change, the effects of the climate on the biosphere and people receded further into the background. Parallel to this "scientification" of climate research, three particular developments emerged:

1. The expansion of our knowledge about climatic relationships of the earth extends into the future as well as into the past. The long-reigning idea in the past century that climate was essentially unchanging gives way to the recognition that clear climatic changes also occurred in historic times.[2] This insight, together with the analysis of factors influencing the climate system, leads directly to the understanding that climate can also alter due to human action. As a matter of fact, climate has clearly changed in the last 100 years, and many climate researchers today are convinced it will continue to do so in the future because of the Greenhouse Effect and other anthropogenic factors.
2. The climate system is globally measurable today through the introduction of satellite-based observation systems. Certainly satellite data extend only a limited period, and therefore are of only limited use for inquiry into long-range climatic developments. "Synoptic" representations (detailed descriptions of the real condition, as in weather maps), at least, of the physical condition of the atmosphere are possible. Scholars aspired to this goal in the 18th century through the first global meteorological measurement network of the "*Societas Meteorologica Palatina*" (1780–1792), established by the Mannheim Academy of Science (Germany). Today such networks are routine and essential requirements for daily weather predictions.
3. The mathematizing of physics also led to a mathematizing of meteorology, oceanography and climate research. Atmospheric and oceanographic processes are described by mathematical equations. Before the invention of electronic calculators, these equations could only be solved by drastic simplifications, so that only the principal features could be studied. The availability of electronic calculators permitted the realization of useful climate models, which could

---

[2] We will return in Sec. 4.6 to this issue, when we list various historical cases in which people perceived the climate as changing due to human action.

realistically describe natural processes and their sensitivity with something approximating reality in relation to anthropogenic influences. These climate models act as experimental devices for climate research.[3]

The depth of understanding of climate dynamics made possible by these new methods has recently allowed climate research to enter the limelight of scientific and public interest.

---

[3] For a conceptual treatise dealing with modeling in climate science, refer to Peter Müller and Hans von Storch, *Computer Modelling in Atmospheric and Oceanic Sciences — Building Knowledge* (Springer-Verlag, Berlin–Heidelberg–New York, 2004).
   Indeed, "modeling" is a difficult term, which is very differently understood in different social and scientific quarters. In climate science, "models" are mathematical constructs, which describe the functioning of the full system by combining components that describe the entirety of all significant processes, preferably by using "first principles", such as mass or energy conservation.

3

Climate as Limiting Condition and Resource

The weather — a frequent and natural theme of conversation — influences our everyday life, our behavior, and not least, our well-being. Almost everybody likes to observe and discuss the weather. Perhaps, then, it is not the fever thermometer, but the indoor and outdoor thermometers that are the most common instruments in modern homes.

Added to these casual conversations about the weather is a second related theme — climate. People often complain that "the weather" has become worse, by which they mean the statistics related to weather — therefore, the climate. Storms have allegedly become more frequent or stronger, the weather less predictable, and seasonal distinctions have been obliterated. In answer to these complaints, it is often asserted that it is humans who are about to destroy, or at least harm, the climate and thereby their own foundation of life. We will see in Sec. 4.6 that this supposedly new theme is by no means so new. Earlier generations asked themselves to what extent their activities could have negatively affected climate. They also asked how far climatic conditions have an influence on living conditions.

In this chapter, we will deal with the static or unchanging climate, which certainly brings extremes from time to time. Catastrophic events such as hurricanes,[4] floods, heat waves, storm surges, and droughts regularly take

---

[4] Or as they are called in other parts of the world: typhoons, tropical cyclones, etc.

place; but in spite of these catastrophes, climate is constant. After these "normal" but rare events, weather conditions return again to the usual condition. A 100-year flood occurs on average once every 100 years; if not, then something is not right with the climate, or with the computation methods for a 100-year flood level. If two 100-year floods occur one right after the other, that is still no reason for alarm.

The climate that we experience has an important characteristic: its reliability or normality. This reliability of climate allows us to deal sensibly with its possibilities, contingencies and inclemencies, although modern society is certainly experiencing a certain erosion of this trust. One reads regularly in the media, in connection with weather extremes, that the climate is behaving eratically, and that this leads back to anthropogenic effects. However, one should realize that the normal weather case includes a certain level of "crazy behavior", and that we only have reason for unease if the weather behaves more crazily than usual, or even if it no longer behaves crazily.

In Sec. 3.1, we deal with climate as a daily human experience and as an occasionally limiting factor on the ecosystem and human activities. In this regard, climate is the statistic of locally experienced variations in the weather. These phenomena express themselves in measurements, such as the duration of sunshine, or the amount and frequency of precipitation, air temperature and wind. These measurements are also relevant for the users of climate information, such as forestry, environmental protection, shipping, street and air traffic, or the tourist industry.

For the scientific climate researcher, however, these measurements are of only subsidiary meaning. For the scientist, climate is a complexly structured system in which such diverse components as the ocean, atmosphere, and sea ice work with and against each other (Sec. 3.2). In this context, the air temperature near the ground is likewise uninteresting. In its place, other measurements occupy the foreground: heat transport, stream function, vertical exchange, or absorption and reflection of radiation by the clouds. This view of climate as an interactive system of physically describable components allows the climate researcher to explain why our present climate is as it is, both in the global sense of climate research and in the local sense of practical users of climate information and laypersons.

The scientific reconstructions of the climate system are virtually irrelevant for the general public. But the scientific construction of the climate system competes with the social understanding of the climate (Sec. 3.3). Neither is independent of the other, but each is relevant and valid in different societal spheres. The scientific construct deals essentially with the formation of the climate: Why are there storms? The social construct orients itself more towards the perceived effects of climate.

In Sec. 3.4 we will discuss an important and still vital school of thought, which continues to inform the social construct of climate, whereby human activity is co-determined in large part by the particularities of the local climate. This school of thought is called Climate Determinism.

3.1. Climate as an Environmental Experience

People experience climate only as a totality of weather occurrences in the places they live; that is, climate is perceived as "typical weather". Part of this typical weather consists of occasional very warm and too dry summers, while other summers are ruined by rain; sometimes many severe wind storms crop up in succession, while in some winter seasons almost no storms occur. "Typical weather" must not become confused with "average weather"; the last is a mathematical artifact, which does not occur in reality. The first includes occurrences of extreme incidences with a certain characteristic frequency.

One of the prominent meteorologists of the late 19th and early 20th century, the Viennese professor Julius von Hann (1839–1921), explained in his Handbook of Climatology,[5] considered by his contemporaries to represent the state of the art:

> "... *Climate science will ... have the task of acquainting us with the average atmospheric conditions over the different parts of the earth's surface.*"

Thus, climatology was considered meteorology's bookkeeper.

---

[5] J. Hann, Handbook of Climatology, Vol. 1: General Climatology (New York, Macmillan, 1903).

Meteorology is something other than climatology, in that meteorology deals primarily with the physics of atmospheric processes. For lay people the main task of meteorology is weather forecasting. For a long time this task was pursued with methods that are today considered obscure (for example, through classifying weather conditions, or through selecting "analogous" situations in the past). Only since the advent of electronic computers at the end of the 1940s has forecasting assumed a scientifically solid and reliable position.

The most important everyday experiences relate to the diurnal and annual cycles; in the morning, before sunrise, it is the coldest and a maximum of humidity condenses. Figure 1 shows as an example the diurnal cycles for summer conditions in Germany. Maximum temperatures are reached at about 14:00 and minimum temperatures not before 6:00. The temperature contrast in the coastal resort of Warnemünde is only about 5°C, and thus much smaller than at inland Potsdam (near Berlin), with a daily temperature range of about 9°C.

The annual course of temperature, the rise and fall of air temperature from month to month, leads to the differentiation of seasons. For a number of stations throughout the world, the mean annual cycles of rainfall, of mean day and mean night temperatures are given in Fig. 2.

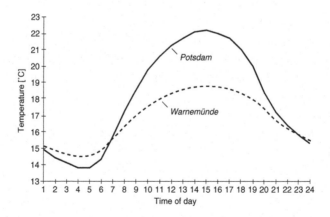

Fig. 1. Mean diurnal cycle of temperature at two stations in Germany in July — Warnemünde at the Baltic Sea coast and Potsdam, close city of Berlin.

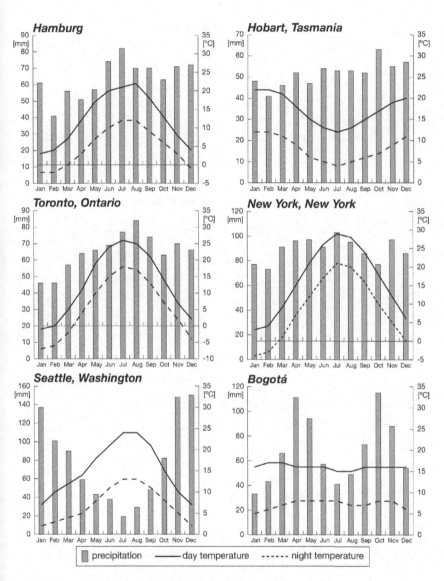

Fig. 2(a). Annual cycles of precipitation, day and night temperature at some stations.

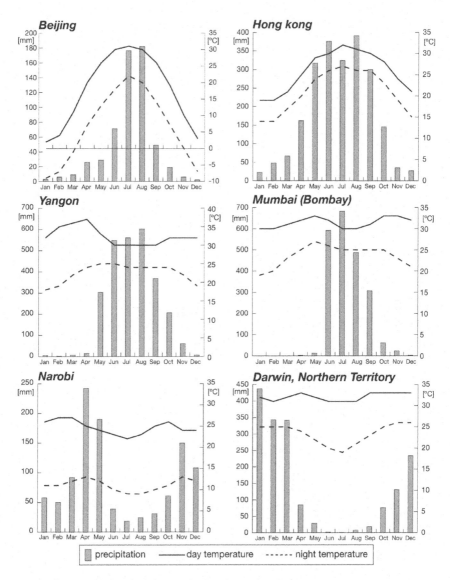

Fig. 2(b). Annual cycles of precipitation, day and night temperature at some stations.

The coldest and warmest months are fixed in our consciousness as winter and summer. Actually, our "official" seasons are determined astronomically. This determination accords well with a meteorological definition, although it makes more sense meteorologically to designate the months December–February as (northern) winter, March–May as (northern) spring, and so forth. By the word "winter" we refer to the winter of the northern hemisphere, because the months December–February are in the middle of the three warmest months of the year in the southern hemisphere.

The division of the seasons, as far as it is based on the observations of the temperature range, is not identical with the length, especially the shortness, of the days. In the earth's temperate climate zones the greatest cold is not recorded on the shortest days, nor the greatest warmth on the longest day of the year. The difference is physically easy to understand: the highest water temperature in a bathtub will not be reached when the hottest water keeps running, but rather when the running water is slightly warmer than the contents of the bathtub (if we disregard other processes).

The reverse evolution of the northern and southern hemisphere extratropics is clearly recognizable in Figs. 2(a) and 2(b), for instance in the case of Hamburg in Germany and Hobart in Australia; as is also the absence of a pronounced annual cycle at the tropical station at Darwin in northern Australia or Bogotá in Northern South America.

In the temperate climate zones of the northern and southern hemispheres, we can separate four seasons more or less clearly from one another. In many areas of the tropics there is a semiannual oscillation, with two annual minima and two maxima instead of an annual oscillation with one annual minimum and one maximum. Examples are Mumbai in India, Yangon in Myanmar or Darwin in Australia. This is caused by the sun, which passes over tropical latitudes and sheds a maximum of radiation twice a year.

Figures 2(a) and 2(b) also show the mean annual course of precipitation amounts, which varies strongly among the stations shown. Some show a monsoonal characteristic with a marked dry period and a marked wet period, e.g., Mumbai or Yangon; others show an almost uniform precipitation amount, e.g., Hamburg, Hobart or New York. A bimodal distribution, with two maxima and minima, can also be found among the cases, namely Nairobi in East Africa and Bogotá.

In temperate climate zones, seasonality is an important human experience. The rhythm of the seasons will be perceived most frequently as positive in everyday life. People from temperate zones who live in other climate zones of the earth, without distinct seasons, will experience the absence of seasons to be an environmental deficit.

The initial "scientification" of climate understanding at the end of the 19th century led first of all to the replacement of undetermined climatic observations such as: "The winter climate of our region is severe" or "Our summers are humid and variable". Today numbers derived from objective instruments describe climate. The scientific study of climate led to a reliable representation of observable climate variables, and thus to a numerical language.

An early example of a methodically reliable measuring technique and organization was achieved by the already mentioned "Societas Meteorologica Palatina",[6] which produced simultaneously comparable measurements in different places in Europe at the end of the 18th century. Figure 3 shows as an example simultaneous air pressure readings at three locations in Europe, which display the eastward passage of a low pressure system during Christmas of 1775.

From the quantitative approach a significant question arose: Which of the many quantitative measurements have real informational value for society and science? The practical need for limiting the number of observations imposed criteria such as robust measurement unaffected by non-meteorological processes, relevance for applications, and representativeness for an area and a segment of time of interest.

Besides (surface air) temperature and precipitation as the most important (bio-) climatic variables, meteorological services regularly record other variables, such as humidity, wind, cloudiness, and the number of hours of sunshine. Another climatic variable that is taken not from

---

[6] An account of this society is provided by J. A. Kington, The Societas Meteorologica Palatina: An eighteenth-century meteorological society, *Weather* **29** (1964) 416–426. See also: C. Lüdecke, The monastery of Andechs as station in early meteorological observational networks, *Meteorologische Zeitschrift* **6** (1997) 242–248.

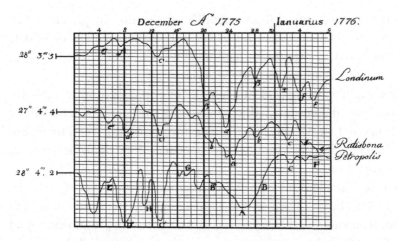

Fig. 3. Air pressure record for London, St. Petersburg and Regensburg collected by the Societas Meteorologica Palatina for the period December 1775 to January 1776.

meteorological but rather from hydrographical services is the water level along coasts and lakes, as well as along rivers.

Climate researchers sought methods to measure the relevant climate variables so that the numerical values could be reproduced for the place in question, and in addition also permit a comparison with other places.

This task sounds easier than it is. Take, for example, the quantity "daily average temperature", which is derived by averaging a few observations throughout the day. The result depends on the actual time when the thermometer is read. Thus, observing not at 6, 12, 18 and 24 o'clock, but instead at 7, 13, 19 and 1 o'clock gives different "daily average temperatures".

That this argument is not an academic one is demonstrated by the fall of ocean surface temperatures by nearly half a degree Celsius in the raw numbers in the early 1940s. This drop was entirely due to a change in observational practice and had nothing to do with the temperature of the ocean. Before, the temperature was recorded by measuring the temperature of seawater hauled to the deck in a bucket; afterwards, the temperature of the water used to cool the ship's engines was read.

The history of meteorology and oceanography knows many such "inhomogeneities" of observation data. The variable is not representative for a certain area, but may be only for one point, or does not allow for comparison across time. Some scientific articles do not present actual changes in the climate system, but rather only changes in the kind of data collection, or data preparation and analysis. The problem is worse, of course, in the case of internet information.

In the following, we present a number of examples of such "inhomogeneities" in climate data, which show spurious developments whose source lies not principally in changing meteorological conditions, but rather in changing observational practices and other changing environmental conditions.

A first example deals with the observation of occurrences of strong wind in Hamburg, shown as the number of days with wind strength 7 Beaufort and more within ten years (Fig. 4). The resulting numbers showed more than 90 of these occurrences in the decades before 1951–60; since then, only 10 more occurrences were noted. The reason for this variation has nothing to do with any climatic changes, but rather with the transfer of observation sites from the maritime weather office in the harbor to the airport. The observations are known to be correct, but they are evidently not representative of Hamburg's storm climate. The numbers shown in Fig. 4 are climatologically unusable in this form to answer the standard questions of the kind: "How high is the severe wind risk in Hamburg?" or "Is there a change in the frequency of severe wind occurrences in Hamburg?"

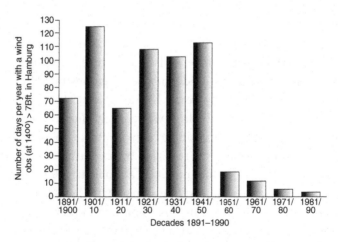

Fig. 4. Frequency of recorded gale wind conditions in Hamburg per decade.

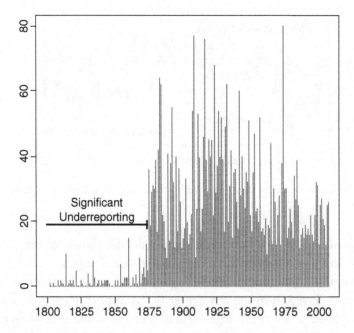

Fig. 5. Frequency of tornado reports in the USA. From Harold Brooks.

A second, similar case refers to tornadoes in the USA (see Fig. 5). Before 1870, reports about the occurrence of tornadoes were sporadic and anecdotal. Only in the 1870s did the Signal Service of the US Army begin systematically to collect reports about tornadoes; but this activity was viewed as politically unwelcome, as immigrants were not to be scared by these violent events. Therefore, tornado activity was described as less significant in the late 1880s; this was only rectified a few years later.

The third case concerns the "urban warming effect". The temperatures observed in cities are higher than those in surrounding rural areas. The air in a city cools less quickly than that in the country because of the reduced evaporation from the widely sealed urban surface of the earth (streets, buildings).[7] Differences of more than one degree appear. This effect is shown in Fig. 6, depicting the annual temperature ranges for two neighboring places in the Province of Quebec (Canada).

---

[7] See W. R. Cotton and R. A. Pielke Sr., Human Impacts on Weather and Climate (ASTeR Press, Ft. Collins, 1992).

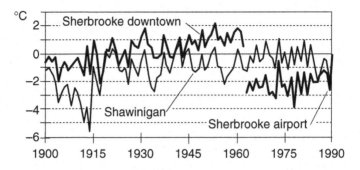

Fig. 6. Temperature record from two neighboring stations near Montreal, Canada.

The weather station "Sherbrooke" represents the climate of the growing city of Sherbrooke, while the station "Shawinigan" represents the rural region around the small city of Shawinigan. In 1966, the Sherbrooke station was transferred from the city center to the airport outside of town. Obviously the same abrupt changes of measurement values ensued on this occasion as already seen with the storm climate of Hamburg. The "Sherbrooke" station is not representative of the city area of Sherbrooke, and especially not that of the city's surrounding region. Furthermore, in the city we see a constant warming that was not measured at the rural weather station. Thus, the weather station "Sherbrooke" is useless for climatological purposes, other than providing a description of what is going on right at the measuring site. The measurements can hardly be used for agricultural planning or for a reliable assessment of whether man-made climate change is going on.

One of the consequences of the systematic "urban warming effect" is that for the determination of area averages, only temperature records of uninfluenced weather stations must be used. Worldwide, there are many long temperature records, some of which go back to the 17th century. But most of the early observations were made in cities like Bologna, some in the earliest times even inside houses. The longest series of observations are therefore only of limited usefulness for reconstructing past climate deviations. This is regrettable, because the assessment of currently observed warming demands a comparison with earlier warming trends caused only by natural processes. For that purpose, reliable, well-documented descriptions of changing temperatures in times without human interference are needed.

The final case refers to a widely used analysis of damages related to hurricane activity. This case featured prominently in the Third Assessment Report of the IPCC, and is a standard argument among activists who push for regulatory political measures to limit greenhouse gas emissions.[8] Figure 7 shows the damage related to hurricanes since the year 1900, when they make landfall on the US coast. The damage is given in 2005 $s. Obviously, we see a dramatic increase in damages, with a peak value in 2005, when Hurricane Katrina hit New Orleans. The case was made that this increase in damages was related to increasing sea surface temperatures

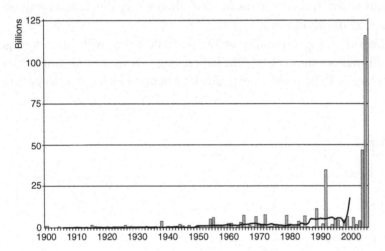

Fig. 7. Annual accumulated damages related to hurricanes making landfall along the US coasts from 1900 to 2005. After Pielke *et al., op. cit.*

---

[8] When we deconstruct this case here, we do not intend to argue that elevated levels of greenhouse gas concentrations in the atmosphere, principally of human sources, would *not* change climate; we also do not want to indicate that such change would not represent a significant danger, or that reduction of greenhouse gases emissions would not be useful or even mandatory. All that we intend to do is to show that this specific argument is mistaken.

See R. A. Pielke, Jr., J. Gratz, C. W. Landsea, D. Collins, M. Saunders and R. Musulin, Normalized Hurricane Damages in the United States: 1900–2005, *Natural Hazards Review* 9 (2008) 29–42.

in the Gulf of Mexico and elsewhere, and that this warming was caused by global warming.

The numbers given in Fig. 7 are acccurate as they are collected by insurance companies. It is the interpretation that is problematic for two reasons; the less significant one is that the hurricane activity shows marked variations from decade to decade. The significant issue is that the usage of the coastal area affected by hurricanes making landfall has undergone massive changes, with many more people living in coastal communities, and much more wealth and capital subject to weather related risks in such coastal communities. When this effect is taken into account, by assuming that the history of hurricanes making landfall since 1900 is unchanged, but the affected wealth and capital along the coast are throughout the 2nd century as in 2005, then a very different assessment emerges, as shown in Fig. 8.

There is strong variability in the past 100 years, with high damage cases throughout the century. The largest single damage event would have taken place in 1926, when a hurricane hit Miami (which was a sleepy little

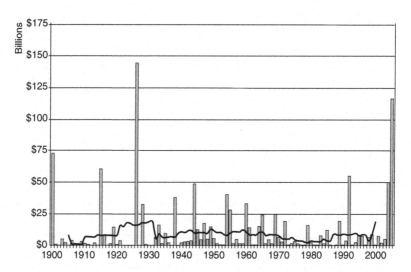

Fig. 8. Same as Fig. 7, but after assuming the same accumulation of people, wealth and capital along the US coast as in 2005, but throughout the entire century. After Pielke *et al.* (2005).

city back then). Katrina cost US$81b, but the "Great Miami" storm in 1926 would have caused damages of approximately US$130b, had Miami then already been a megacity, as it is today.

Figures 7 and 8 convey very different stories. The former informs us that the recent level of hurricane induced damages is unprecedented, and this change is due to a unprecedented level of hurricane activity. One may expect a further increase in the coming years and decades. Figure 8, on the other hand, informs us that since 1992 very heavy damages have been incurred by hurricanes, but the damages were within the range of historical precedents. Figure 8 also demonstrates that 50 years of data, a time interval often used, is insufficient to estimate the full range of possible events.

The point is that all climate variables fluctuate in time as well as in space — from day to day, year to year, century to century and so forth. Wind speed or air temperature changes in a range of seconds, just as it does in the course of weeks, years, or centuries. Obviously, there is a need to quantify the range of variations, by which the climatic variables usually vary and the probability of extreme conditions. Only when we have such measures of "normal" variations can we decide whether we are confronted with anthropogenic changes.[9]

In this situation, it is useful to refer back to the abstract concepts of terminology of statistics. Climate variations are irregular,[10] consistent with random variations, except for the regular above-mentioned annual and diurnal cycles. To be more exact, we view the deviations from the annual and diurnal cycles as a random process. Randomness is a

---

[9] We return to the issue of this "detection" in Sec. 4.4.

[10] When harmonic analysis was invented, which decomposes all series into periodic components, there were many efforts to detect and isolate the many periodical components in weather, as well as in finance and other issues. It took many decades for people to find out that even completely random series can be decomposed in this manner, and that extending the time series by just one more number may destroy the purported periodicity. While the concept is useful when dealing with truly periodic phenomena such as tides, in the climatological context the straightforward application of the methodology leads to artefacts; harmonic analysis, however, is still popular among the naïve.

mathematical construct with whose help we can describe the apparent irregularity well.[11]

In the following we venture into statistics. We want to understand a process that produces a series of unpredictable numbers whose values follow a certain probability distribution. The best known of such distributions is the bell-shaped Gauss or "normal" distribution. Probability distributions describe with what probability possible different values occur. One can describe such distributions by a few characteristic numbers — above all by the mean value and the standard deviation.

The mean value is simply the arithmetical average of all observations. In most cases, one half of all observations are smaller than the mean value and the other half is larger than this number.[12] The annual and diurnal cycles in Figs. 1 and 2 are such mean values (averaged for each calendar month and each hour individually).

The standard deviation is a measurement of the range of random numbers. In most cases, two-thirds of all random numbers will be within the interval "average plus/minus one standard deviation"; but about one-third of all numbers are greater than the "average value plus one standard deviation" or smaller than the "average value minus one standard deviation".

Randomness does not mean that two consecutive numbers are independent of one another; we observe much more often — especially in a climatological connection — that the value of a climate variable at a point in time will be determined in part from the value at the preceding point of time, according to the maxim: "The best forecast for tomorrow's weather is today's weather".

Logically, the value at any time in future is then also partially determined by the actual value; but the degree of determination decreases with

---

[11] Here, it does not really matter whether we are dealing with "real" randomness, whether "God is rolling dice". It suffices to note that the multitude of non-linear, often chaotic processes in the climate system together generate a time behavior which cannot be distinguished from the mathematical construct of randomness. Thus "randomness" is a convenient, and indeed powerful, tool for conceptualizing climate variability. See also Sec. 3.2.

[12] Strictly speaking, this is valid only for symmetrical distributions.

time, so that after a sufficiently long time the future values and the actual value have practically nothing more to do with one another. Here, this "having nothing to do with one another" means that a random reshuffling of the series does not change the character of the series. On the other hand, when the character of the series is changed after reshuffling its elements, then some kind of "memory" prevails in the series. The stronger the degree of determination from one number to the next, the larger the memory.

In practice, neither the distribution nor the memory is known. Therefore, one has to estimate these characteristic numbers from observations. How many observations are needed in order to do this sensibly? One can record the temperature for twenty years and compute the average value for the first and the last ten years separately. Both of these averages will be different. In order for the ten-year averages to be representative, the difference must not be large.

Because there are no "natural" definitions for the averaging time, meteorological services have agreed on certain conventions and a standard for climatological averages. This standard amounts to an observational period of 30 years. The world's meteorological services are requested to adhere to this standard. In scientific climate research this standard hardly plays such a role, once it becomes clear that climate also shows notable variations for time scales of 30 or more years.

We can calculate characteristic numbers for diverse climate variables and measuring sites, and offer them in the form of maps to the scientific community and the public as climate information. Figure 9 shows two maps with the tracks and intensity of East Asian typhoons in two years: 1993 with a maximum of cyclones, and 1998 with a minimum. The cyclones form in the West over the North Pacific. Most of them then move westward towards the Asian coastlines. Obviously, the number of events varies strongly from year to year.

A second map deals with excessive wind speeds in Northern Germany. Shown is the peak wind speed, averaged over 2 seconds, which is expected to be observed once in 50 years (Fig. 10). According to this map, right along the coastline one expects peak wind speeds of 50 m/s, while further inland the peak wind speed, occurring once in 50 years, is 38 m/s.

This information is needed in planning and preparing economic and political decisions whenever climatological risks, or benefits, are present.

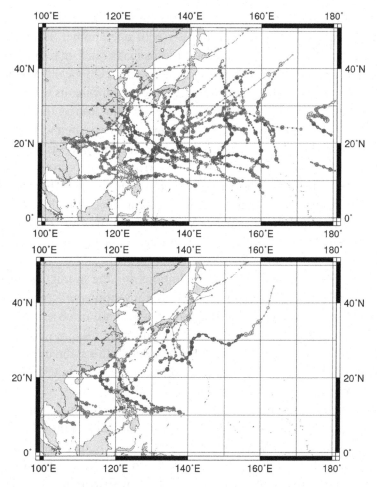

Fig. 9. Tracks of East Asian typhoons in two years: 1994 with a particularly large number of cases (36 typhoons), and 1998 with very few (16 typhoons). The colors indicate different strengths of cyclones.

Source: http://agora.ex.nii.ac.jp/digital-typhoon

Such deliberations encompass a broad thematic spectrum. The viewpoint is almost always anthropocentric. The following topics are typically covered:

(1) The possible effects of climatic conditions on the life of individuals, their well-being and health. We will return to this point.

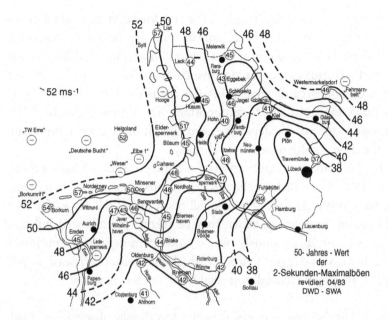

Fig. 10. Map of estimated maximum wind gusts, which take place on average once in 50 years. The area shown is the southeastern part of the North Sea.
Source: German Weather Service.

(2) Adaptation to dangers related to meteorological extremes is another important application for climate information. A typical case is the danger from storm surges and flooding along coasts and rivers. Statistics of precipitation amounts and storm intensities are decisive indicators for estimating and assessing the flooding hazard and the required height of dikes. This group of climatologically determined dangers also includes landslides and mudflows. The frequency of such events is clearly coupled with precipitation statistics; however, as with volcanic eruptions, the individual events are hardly predictable for several days in advance.

(3) Climate statistics are not only significant for individuals and society, but also for the plant world. What type of agriculture will be economically feasible depends not only on the average summer or winter temperatures, but also on the most extreme cold or the timing of the beginning of the frost period. Once an extreme cold temperature has

occurred, it is as a rule largely irrelevant for vegetation whether this temperature is observed repeatedly later on or whether, in spite of the initial extreme, the entire winter period on average qualifies as "normal". In Florida's case, the decisive circumstance is whether there is frost or not, because frost ruins the citrus crop. In other cases the extreme temperatures are unimportant. The timing of the bloom of European snowdrops depends essentially on the average temperature in the months of January and February.

(4) A further important application of climate statistics is whether events are related to anomalous climatic conditions or caused by other, non-climatic processes. Into this class fall questions of the causes for algae blooms in coastal seas, which may be related to enhanced influx of nutrients or anomalously warm and calm weather conditions. Another case is damage to forests, which may reflect increased air pollution, or may be related to several harsh winters in a row.

The industrious inventory of the countless climate observations made for more than 100 years provides important working material for basic and applied climate research. Examples relate to the teleconnections that make regional climate anomalies felt in other regions far away. A prominent case is the El Niño/Southern Oscillation phenomenon, which was described at the end of the 19th century by the Swede H. H. Hildebrandson. Another large-scale climate anomaly is the North Atlantic Oscillation, which describes an anti-correlation of atmospheric pressure and the temperature in the area of the North Atlantic. If the temperatures over Greenland are higher than normal, then, as a rule, they are lower than normal over Northern Europe, and vice versa. Coupled with that "seesaw" is a raised atmospheric pressure over Iceland and a reduced atmospheric pressure over the Azores, and vice versa. This mechanism, which was presumably described for the first time by the Danish missionary Hans Egede Saabye in his publication entitled *Dagbog holden i Grønland i Aarene* 1770–1778, is of great importance for the European climate.

The number and relative significance of climate variables has changed in the course of scientific concern with the climate. In the past, one used to proceed with an analysis of individual climate variables in an isolated manner. Today, however, one tries to include diverse climate variables that are used to describe an integrated climate system. The purpose is to better understand the functioning of the climate system as a whole, and so one must include such factors as the oceans, sea ice, the biosphere, and so forth.

A hundred years ago, inadequate technical capabilities limited observations essentially to the surface of the earth, thus restricting the definitions and borders of climate research. But in the 1920s it became possible to obtain observation data from different altitudes, with the help of weather balloons, kites, aircraft and radio soundings. This opportunity enabled the discovery of the stratosphere at the beginning of this century. From observations during mountain climbs and spectacular manned balloon flights, it was learned that the temperature sank about 0.7°C per 100 m of altitude. From that information Hermann von Helmholtz (1821–1894) concluded that at about 30 km, absolute zero (−273°C) must be reached. When the first measurements from unmanned balloons revealed a constant temperature above 11 km, many meteorologists at first believed that errors in the measurement had been made. Actually, it was the transition from the troposphere to the stratosphere.

These new types of observations have effected a fundamental change in the science of climatology. After several decades it is no longer a geographical sub-discipline, but rather a kind of environmental physics and chemistry. It is not surprising that young meteorologists in particular were fascinated with this development and accelerated this paradigm shift in meteorology.[13] In the following section, the understanding of this "new" climate research will be more closely explained.

3.2. Climate as a Scientific System

In order to demonstrate the difference between the descriptive, geographical, tradition-bound climatology and physically oriented climate research, we want to deal first with the Greenhouse Theory of the Swedish chemist

---

[13] The following book provides an interesting account of the transformation from the purely descriptive approach towards dynamic thinking: R. M. Friedman, Appropriating the Weather. Vilhelm Bjerknes and the construction of a modern meteorology (Cornell University Press, 1989), p. 251.

Svante Arrhenius (1859–1927) as an example of a typical "physical approach". At the end of the 19th century, a question much discussed among physicists and chemists was: Which factors determine the temperature of the near-surface air? Behind the question was the recognition that thousands of years ago there was an Ice Age, so this temperature must have displayed considerable deviations. Arrhenius, who later received the Nobel Prize in chemistry for other achievements, postulated that the incoming short-wave (solar) radiation would be balanced by the outgoing long-wave (thermal) radiation emitted from the earth. If it did not balance, the temperature would fall or rise until this balance was reached.

Were a vacuum to exist between the sun and the earth, the average air temperature of the earth would amount to about −10°C. This is apparently not the case. Between the sun and the surface of the earth is the earth's atmosphere. As the radiation passes through the atmosphere, it encounters several obstacles, namely gases such as water vapor, carbon dioxide or methane, small particles and clouds. Some of these gases are "greenhouse gases" that catch long-wave radiation, which originates from the surface of the Earth and from the atmosphere itself, and radiate it back in all directions. Thus part of the energy radiated from the ground does not escape directly to space, as would be the case without an atmosphere, but is partially returned to the ground and lower levels of the atmosphere after having been absorbed and re-emitted by the greenhouse gases. Greenhouse gases already work in this way at very low concentrations. Water vapor is the most abundant and efficient greenhouse gas, while carbon dioxide makes up only 0.03% (volume percent) of the air.

Only about 40% of heat radiation "gets through" to space, while 60% of the energy is radiated back. Not only the short-wave solar radiation, but also the long-wave heat radiation returned from the atmosphere arrives at the ground. If we assume that our system normally has the above-mentioned temperature of −10°C, then it would indeed warm up as it gathers energy. The warming causes an increase in intensity of the long-wave radiation; of which, again because of the greenhouse gases, only 40% continuously reaches space. Because the radiation grows with the temperature, the temperature increase causes more long-wave energy to be emitted. The warming exhausts itself when the 40% of the radiated long-wave energy that reaches space compensates for the solar radiation arriving at the earth's surface. A "final temperature" appears that is considerably higher than the original −10°C. However, because the atmosphere not only absorbs and re-emits long wave radiation, it also shields the ground to some extent against incoming solar radiation, actually not 100% of the solar radiation found at the top of the atmosphere reaches the ground, but only a portion. The shielding depends on the albedo,[14] which is related to clouds, sea ice, snow cover, desert and land use. The combined effect is moderated so that finally an average temperature of about 15°C appears, which agrees with observation.[15]

This is the "Greenhouse Theory", actually a confusing designation because the temperature in a vegetable greenhouse is, for other reasons, warmer than that of the surrounding air. Noteworthy about this theory is the fact that today, 100 years after its publication, it is still acknowledged in almost unchanged form as correct.[16] Svante Arrhenius's reflections also showed that alterations of the carbon dioxide concentration in the atmosphere could in fact account for the origin of the Ice Age. Investigations of ice cores have revealed significant parallel variations of carbon dioxide concentrations and temperature (Vostok ice core). Whether the enhanced greenhouse gas concentrations were really a cause for the Ice Ages, or but a consequence of changed climatic conditions, we cannot say for certain. Also, in the meantime, other plausible hypotheses for the origin of the Ice Age cycles have been proposed, mainly related to the changing geometrical arrangement of the sun and the earth.

Svante Arrhenius also calculated the rise in air temperature in the case of a doubling of the atmospheric carbon dioxide concentration, and found a value comparable with present estimates of about 3°C. He maintained that a doubling was certainly possible, but only in 1,000 or more years,

---

[14] This is the property of a surface, such as that of the earth or the sea, to reflect short-wave radiation, measured in percent. Deserts have a high albedo, snow even more, but forests have a low albedo. The albedo of newly fallen snow is up to 95%, while over the sea it can be less than 10%.

[15] In fact, this representation is simplified. A number of other processes, such as convection, modify the picture.

[16] S. A. Arrhenius, On the influence of carbonic acid in the air upon the temperature of the ground, *Philosophical Magazine and Journal of Science* **41** (1896) 237–276. The original publication was in German.

because 85% of the carbon dioxide emitted into the atmosphere would accumulate in the ocean.[17] Today we know that the absorption in the ocean takes place over a protracted period of time, and therefore a doubling of the CO_2 concentration in several decades is quite possible, and even very probable. We will return to the question of anthropogenic climate change in Chap. 4.

The purpose of climate studies is no longer to collect and analyze many detailed observations in order to facilitate various planning objectives. Instead, the research strategy exploits fundamental physical principles, among them the principle of energy conservation, the first law of thermodynamics. Climate becomes an object of basic scientific research. The role and meaning of the observations are mainly reduced to the role of "validation" of models and to the "refutation" of hypotheses and theories. Even though this type of climate science constitutes fundamental research, it has achieved overarching practical importance, as it is directly influencing international policy. More is mentioned in Chap. 4.

Other notable efforts that aim to explain the general atmospheric circulation were initiated by the Englishman George Hadley (1685–1768) in the 17th century. Even though only very little empirical information was available, and in particular no data beyond the boundary layer of the atmosphere, he grasped essential parts of the general circulation correctly (Fig. 11), such as the trade wind system. But he could not deduce other significant parts at that time.

The German philosopher Immanuel Kant of Königsberg (1724–1804) also worked in this field of study; he analyzed wind observations from ships in Southeast Asia, drawing from them the conclusion that there must be a continent further in the South, which was at that time still unknown Australia.

---

[17] See the remarkable textbook: S. A. Arrhenius, *Das Werden der Welten* (Leipzig Akademische Verlagsanstalt m.b.H., 1908). An English version, *Worlds in the Making: The Evolution of the Universe*, is available in paperback. In this book, Arrhenius describes many aspects of the climate system correctly and comprehensibly — but he failed to describe the functioning of the sun properly, as he had no knowledge of nuclear processes driving the sun. Instead, he speculated about obscure chemical processes. This case may serve as a warning to modern scientists to be prepared to find part of one's own explanatory system refuted.

Fig. 11. Description of the global cell structure as given in the 17th century by George Hadley, who had very little data. Compare with the modern description given in Fig. 13.

In the first half of the 20th century, physically-oriented climate research received a fresh impetus through the work of such researchers as the Norwegian Vilhelm Bjerknes,[18] who contributed to the authoritative explanation of the inner structure of storms in the middle latitudes; the Swede Carl Gustav Rossby (1898–1957), who explained the instability of the weather in the middle latitudes; and finally the American John von Neumann (1903–1957), who after the Second World War recognized the possibilities of electronic data calculations for weather predictions in particular, and implemented the first applications of the newly developed computers.

In the current climate paradigm, the climate system is understood as an interactive system made up of many mutually influential processes in the atmosphere, hydrosphere, cryosphere (the sphere of ice and snow) and

---

[18] Vilhelm Friman Koren Bjerknes (1862–1951); see Fig. 12 and consult Friedmann, *op. cit.*

Fig. 12. The Norwegian meteorologist and founder of the polar front theory Vilhelm Bjerknes, as shown in a painting by Rolv Groven. The painting is located in the Geofysisk Institutt, Universitetet i Bergen.

biosphere, and not as a process that essentially restricts itself to the lower atmosphere. This kind of investigation is no longer descriptive, but rather primarily a systematic analytical process. To first-order approximation, the system can be described with the help of an internal combustion engine driven by the temperature contrasts between the combustion boiler and cooler. In the case of the atmosphere, the tropical convection, which sets energy free by condensing water vapor, is the "active element", while in the case of the ocean the sub-polar "cooler" maintains the global circulation.

The modern understanding of the general circulation of the atmosphere is depicted in Fig. 13.

The heating of the atmosphere successfully takes place by means of the interception of short-wave solar radiation, especially in the tropics. The air near the earth's surface becomes strongly warmed and the air layer columns become unstable; low-lying air becomes lighter than the air above. In this manner, vertical air transport movements are formed. These

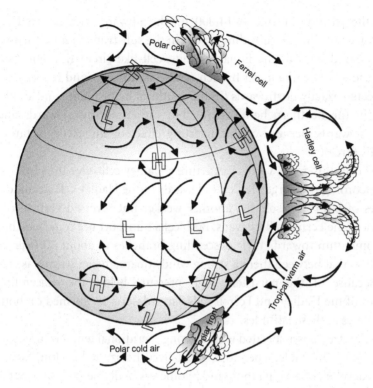

Fig. 13. Modern sketch of the general circulation of the global atmosphere.[19]

are amplified by the fact that when air ascends, it expands, cools off and can therefore retain less water vapor, so that a part of the gaseous water becomes fluid again. During this condensation, the energy originally needed to evaporate the fluid water into water vapor is released as heat. Therefore, water vapor is thought of as "latent" heat. This additional energy warms the rising air, which thereby becomes lighter again than its surroundings and can therefore continue its ascent. If one is underway in an airplane in the tropics, one can observe this process well in the powerful cloud towers that often exceed 11,000 or more meters, often above the flight altitude.

---

[19] See H. von Storch, S. Güss and M. Heimann, *Das Klimasystem und seine Modellierung. Eine Einführung* (Springer-Verlag, 1999).

At the altitude of 10,000–14,000 m, the upwardly transported air flows polewards, finally subsiding slowly in the sub-tropics and returning to the ground towards the Equator. Thus, cell-like circulation emerges, with trade winds as the lower branch. These stationary wind regimes are not directed exactly northwards (in the southern hemisphere) and southwards (in the northern hemisphere), but rather northwest-wards and southwest-wards, because of the diverting effect of the earth's rotation (Coriolis force).

One calls this entire vertical structure a Hadley cell, after the physicist and meteorologist George Hadley. Polewards of the Hadley cell, at middle latitudes of both hemispheres, another weaker but reversed vertical cell circulation, the Ferrell cell, prevails. Both cells transport not only heat, but also momentum towards their descending branches at about 30 degrees latitude; so that between them a strong wind stream, the jet stream, is created. Because of the persistent high pressure below the descending branches of the Hadley and Ferrell cells, large deserts are formed on both hemispheres at these latitudes.

The jet stream is associated with a strong meridional temperature contrast, which is unstable. Therefore, short-lived horizontal cyclonic wind systems develop, of about a thousand kilometers in diameter — the extratropical storms, which transport heat and water vapor polewards.

Thus, one may think of an air particle which travels from the heating areas near the equator to the cooling areas near the poles. At the beginning of this journey more energy is intercepted than is lost, but the farther we go poleward, the less solar radiation arrives, until finally the energy balance becomes negative; the system loses more energy than it receives. The difference is counteracted by the heat transport due to the winds and the ocean currents. Thus, the wind is caused by the spatial differences in the atmospheric energy balance (net profits in the tropical latitudes; net losses in the polar latitudes). The movement originates just as in the case of a steam locomotive — there the back and forth movement of the piston rod, here the wind — from the thermal difference of boiler and cooler, of the tropics and polar areas.

In principle, the circulations of the southern and northern hemispheres of the earth resemble each other. Because of the larger land masses in the northern hemisphere, a disproportionate thermal contrast in the east-west direction takes place. In summer the land masses warm faster than the thermally sluggish ocean, and in winter the ocean cools off more slowly. The consequences of this disproportion are not only the monsoon wind systems in the tropics, but also large-scale enduring meteorological differences in the east-west direction in the Northern Hemisphere. Also, the mountains of the northern hemisphere, the Himalayas, Rocky Mountains, and Greenland, further expedite these east-west structures; the European mountains, including the Alps, play only a regional role.

In the Southern Hemisphere one finds no distinct east-west symmetries. The circulation displays much more than the above-sketched structure of the unstable jet stream with associated storms. Because of these permanently present storms at 40–50 degrees south, the middle latitudes in the southern hemisphere are called the "roaring forties". If one looks at a map of monthly or annual mean air pressure in the southern hemisphere, one finds isobars (lines of equal air pressure) mostly parallel to the circles of latitude. But a look at the daily weather maps shows that the daily pressure distribution is in no way characterized by such a regular pattern. Instead, there are always four to seven storms in a latitude belt over the southern ocean in the roaring forties. Because of the storms, with their regional circular isobars occurring overall in the middle latitudes, the isobars in a map of a time average pressure are smooth and free of regional details.

The oceanic circulation is driven by two mechanisms: by the thrust of winds blowing above the ocean's surface and by subsiding sea water in sub-polar regions, due to cooling sea water and ocean sea ice formation. The circulation in the upper layers of the ocean is essentially driven by the wind. The wind is not only responsible for the upwelling of plankton-rich coastal seas along the South and North American west coasts, but also for the Gulf Stream in the Atlantic and the Kuroshio in the Pacific.[20]

---

[20] The physics of this circulation is explained for non-specialists in a wonderfully clear way in the book *A View of the Sea: A Discussion between a Chief Engineer and an Oceanographer about the Machinery of the Ocean Circulation*, by the oceanographer Henry Stommel, 1991. Stommel cloaks the explanation in a discussion between an oceanographer and a ship's engineer who are together on a research trip.

The circulation of the "deep ocean", that is, the ocean currents at the depth of thousands of meters, is maintained by the difference in density in the vertical sections at different levels of the ocean. In principle the same thing happens in the atmosphere, except that the driving is not provided by warming from below (in the tropics), but rather by cooling from above, at the surface of the sub-polar oceans. This cooling makes the water heavier ("thermal effect"). Also, the formation of sea ice makes the water heavier, because salt is ejected during that process, so that the water becomes more salty ("haline effect"). Thus both effects, the thermal and the haline effect, make the water heavier. If the water at the surface of the ocean becomes heavy enough, the vertical water column becomes unstable and convection sets in. Water from the surface sinks into the deep. This reaction occurs in today's climate in the northern North Atlantic and in the southern ocean at the margins of Antarctica. "Deep" water is displaced away from the convective area. In other areas, such as the Pacific Ocean, the water finally rises to the surface again on a broad scale. One calls this circulation thermohaline.

The thermohaline circulation is much slower than the wind-driven circulation, so that it is less important for the conditions in the upper levels of the ocean; but it determines the conditions of the deep ocean, and thereby also the climate on the surface, for long periods. The present cold conditions of the deep ocean — the water temperature near the ocean floor is close to the freezing point — is by no means necessary or normal. As the American Thomas Chalm Chamberlin (1843–1928) showed in 1907, the deep ocean was warm in earlier geological times.[21] A complete cycle on the "conveyor belt" of the global thermohaline circulation needs 1,000 to 2,000 years. The water that exists today on the floor of the Atlantic began its journey there from the surface in Viking times. The slow subsiding of the water in the depths of the ocean can be reconstructed very well by following the intrusion of radiocarbon (^{14}C) and substances released into the atmosphere during the nuclear tests in the 1960s.

The ocean is no passive component in climate, merely reacting to events in the atmosphere. It influences the atmosphere strongly, in that it

---

[21] See T. van Andel, *New views on an old planet. A history of global change*, 2nd edn. (Cambridge University Press, 1994).

provides something like a heat bath, with moderate temperatures at the bottom of the atmosphere. Also, the ocean, comprising 71% of the earth's surface, enables the generation of a global water cycle, which makes life on earth possible. From the surface of the ocean, water evaporates into the atmosphere. This water vapor influences the atmosphere's radiation properties, and thus the amount of energy that the atmosphere receives from the sun and that is returned to space. Where water vapor condenses, thermal energy is released. The condensed water vapor falls as rain or snow to the ground, penetrates into the ground and runs through rivers back to the sea, completing the global water cycle.

The cryosphere consists of ice and snow areas. It has two functions in the dynamics of the climate machinery: (1) it isolates the ocean and the soil from the atmosphere, so that the exchange of heat and moisture will be reduced and (2) ice and snow surfaces reflect short-wave solar radiation more efficiently, i.e., have a higher albedo, than other surfaces, such as the ocean, deserts and vegetated surfaces.

The earth's atmosphere is therefore no isolated physical system, but rather stands in multiple cause-and-effect relations with the other spheres of the earth system.

As already mentioned, the dynamics of climate generate variations on all time scales. The dynamic mechanism for this generation is different for different phenomena. Apart from the regular variations of the annual and diurnal cycles already mentioned, this variability is largely due to internal processes. The crucial catchwords are "nonlinearity" and "infinitely many dynamic factors". The first effect is the "Butterfly Effect"; the flap of a butterfly's wing can radically change the development of the entire system. Small causes may quickly initiate large effects. This does not mean that a totally different state emerges, but rather, insignificant disturbances may cause a different, but equally likely, sequence of events to evolve. With or without "butterfly flaps", next year's typhoon season will see typhoons in East Asia; but it may depend on the "butterfly" as to whether it will be more similar to the 1994 season with many cyclones, or to the 1998 season with a few (Fig. 9). The second effect can be metaphorically understood as the existence of millions of butterflies, beating their wings uninterruptedly; the effect of all these wings beating is irregular and appears accidental. The dynamic of the climate system transforms part of this apparent randomness into a structured large-scale pattern of variation.

The gravitational forces of the sun and the moon belong to the external factors that cause significant variations in the earth system, namely the oceanic and atmospheric tides. Other factors are the changing radiation of the sun, alterations in the optical properties of the stratosphere due to volcanic eruptions, changes in the parameter of the earth's orbit and changes in the position and topography of the continents. The effect of volcanic eruptions is usually limited to one or two years. The extent of the effect of the sun's radiative output is disputed. The orbit parameters and the topography vary on time scales of thousands to millions of years.

Finally, we want to discuss the relationship between global and regional and local climate.[22] In the classical tradition, one infers knowledge of the global climate from the knowledge of all regional climates. In the scientific sense, this approach does not prove adequate. As we have seen, the different radiation regimes in high and low latitudes determine the global atmospheric structure, with the Hadley Cell in the tropics and the westerly wind and storm zones in the middle latitudes, modified by the existence of great mountain ranges and the distribution of land and sea. Actually, only the very largest structures are of importance for the shaping of the global climate; the disappearance of the Australian continent would not significantly affect global climate, at least in the calculation of a climate model (but naturally, it would change the climate of the Australian region).

On the other hand, the regional climate is the global climate modified by regional details: that is, land use (deserts, tropical rain forests, steppes), regional mountain ranges (e.g., the Alps), and marginal oceans (e.g., the Mediterranean Sea) and large lakes (e.g., the Great Lakes).

Local climates originate out of regional climates through adaptation to such local details as large cities, small lakes (e.g., Lake Constance) and small mountain ranges (e.g., the Appalachian Mountains). That this view of the formation of climate is adequate is demonstrated by the success of climate models that, depending on the degree of complexity, only show structures that describe features with an extent of over at least many

---

[22] See H. von Storch, The global and regional climate system, in H. von Storch and G. Flöser: *Anthropogenic Climate Change* (Springer-Verlag, 1999).

hundred, or even a thousand, kilometers. In such models, no local or small-scale features are described out of which realistic local or small-scale climates could be constructed. As a rule, the regional climates are inadequately simulated. Independently of these shortcomings on local and regional scales, these models are successful in simulating the global climate.

Finally we want to briefly return to the manifestation of climate, the weather. Ultimately, the weather is the reason why we are interested in climate, because climate gives us a measure of the reliability of weather, a framework for arranging our lives.

The temporal development of the weather as it appears on the weather maps, particularly the development, migration and disintegration of high- and low-pressure systems, differs from the externally determined cycles of the annual and diurnal cycle. Due to its highly nonlinear dynamics, the weather is predictable only for the approximate life spans of the low- and high-pressure areas, i.e., for a few days. The difficulty in making predictions grows with the instability of large weather regimes. While the arrival of a storm in the tropics as well as in the extratropics may be predicted a few days in advance, predictions for the arrival of smaller, short-lived systems, such as rain bands or thunderstorm zones, are only possible for hours, or a day, in advance.

This limited predictability of the weather is not a contradiction to claims that one would be able to predict, or to describe in scenarios, the climate many years into the future. If we know how the statistical properties of the random process "weather" change, then we know what to expect as possible, but not what the concrete weather will be.

3.3. Climate as a Social Construct

Weather and climate have been of great importance from time immemorial. Conversations about the weather occupy a prominent position in everyday life, and indispositions of every kind are cheerfully or seriously attributed to the weather. No modern mass medium could give up its regular weather predictions. The analysis of the climate, especially its effect on people and society, has always exerted a fascination for people and for science.

The well-known German physician, the founder of modern pathology, and anthropologist Rudolf Virchow (1821–1902) spoke to a gathering of scientists more than 100 years ago about the problem of acclimatization:

> "We know that a person who goes out of his fatherland into another country which is markedly different ... perhaps experiences in the first moment a certain animated renewal, but after a short time, mostly after only a few days, feels somewhat uncomfortable, and that he requires a few days, weeks, even months depending on circumstances to find his equilibrium again ... that is something so generally known that every man knows and expects it; one assumes that everyone who arrives in such a country and is not completely negligent uses precautionary measures in order to ease this period."

Virchow moreover asserts that the human organs literally alter themselves in this phase of acclimatizing. This process, as he says, is something like a permanent redressing. According to Virchow, this process may even lead to a climate sickness. The newly formed, climatically adapted organs, according to Virchow's conjecture, may become permanent, so that they are bequeathed to descendants.

There were historical periods when the occupation with the climate and its influence on people, society or whole sections of civilization, on forms of government, symptoms of illness, truth, morals, etc., formed one the most important questions for science and philosophical discourse. These various efforts and their modern echoes will be elaborated upon in a later section.

As Astrid E.J. Ogilvie and Gísli Pálsson point out:

> "Allusions to the creation of fair or foul weather by magical means may be found in many different types of writings, from Homer to Shakespeare. Clearly, the wish to be able to control the weather is a deep-seated human desire. In countries where the weather is capricious and frequently wet and stormy it is not surprising that supposed magical control of the weather has been elevated to a high art."[23]

---

[23] In Scandinavia generally "fishermen and sailors employed the services of wise folk to make favorable sailing wind. But beyond that, people quite commonly made use of spells and rituals to influence the weather". (R. Kvideland and H. K. Sehmsdorf, eds.), *Scandinavian Folk Belief and Legend* (University of Minnesota Press, Minneapolis, 2002).

Both everyday and more systematic concepts of weather, magical and climate were closely linked with religious and astrological perceptions in times when religious views of life dominated. In antiquity the gods were responsible for the weather. One could obtain information from the gods about future weather conditions. Certainly the function of the priests was not only limited to obtaining weather prognoses; different rituals and symbolic actions were designed to influence the gods to send specific weather conditions.

In the Middle Ages, bad spirits were also held responsible for weather and weather extremes. Women were denounced as weather witches and burned at the stake. Weather extremes such as floods and storm surges, drought, and hail; and also their indirect, frequent effects, such as plagues of mice, pestilence, cattle epidemics, and bad harvests, were interpreted as a return to biblical plagues or even as apocalyptic signs. Such occurrences and their social and economic consequences, for example, scarce and expensive food, were regarded by people as no random events, but God's punishment for man's sinful conduct. Witches who committed the worst sins were relentlessly persecuted. The persecution of witches was a type of political climate policy of the time.

There is also the opposite tendency, to conclude that religious belief systems develop out of climatic conditions. The French philosopher Voltaire (1694–1778), for example, was convinced that monotheism originally arose in desert regions. The attempts to explain the variety and peculiarities of religious views of life and beliefs by relating them to climatic factors did not withstand modern scientific inquiry, and have been abandoned today.

Babylonians and Egyptians tried to formulate weather forecasts from astronomical constellations. Greek philosophers continued this art. Astronomical weather prophecy was just as widespread in the classical Roman Empire. Moreover, astrological views were combined with a geocentric view of the world where everything beyond the earth is related to the earth and exists for the earth. The seven planets then known were like weather sovereigns, who decided in their own ways about the weather on earth. Saturn, the principal planet, was responsible for cold and damp weather conditions. Mercury was responsible for cold and dryness, while the sun, of course, was responsible for warmth. Each day of the week, and each year as well, was ruled by one of the seven planets. It was therefore relatively easy to predetermine which kind of weather would dominate a certain year: one need only divide the year's number by seven. 1996 divided by seven gives the remainder of one. Therefore, the first planet, the sun, would determine the weather of this year. 2010 should therefore be a warm and dry year. On the other hand, 2011 would be influenced by Venus, and would therefore be more moist.

Astrology has by no means disappeared and lost its appeal. It still lives and has significance for many people, even today, in making weather forecasts. Even the natural philosopher and astronomer Johannes Kepler (1571–1630) could not withstand the temptation of astrology. He said:

"*Astrology is astronomy's foolish darling daughter; but she supports her mother.*"

The hundred-year calendar (first printed in 1700 in Erfurt, Germany with its combination of weather observations and folk sayings) continues to sell quite well today, and finds wide use and acceptance as a means of predicting the weather, although in this respect it is completely useless.

3.4. Society and Humans as a Climate Construct

The discussion of climate consequences on humans and society in the classical period includes the ideas of Greek and Roman philosophers, especially Hippocrates, Plato and Aristotle, as well as observations in the early Middle Ages and the Renaissance.

The writings of the ancient Greek physician Hippocrates of Kos (ca. 460 to approximately 370 B.C.), above all his theses about the effect of food, occupation and especially climate in causing diseases, which regained influence in the Middle Ages, the Renaissance and in the period of the Enlightenment,[24] play an important role. His book *Airs, Waters and*

---

[24] French revolutionary physicians revered Hippocrates and employed his ideas to promote a secular utopia. The philosophical inclined doctors, the ideologues, were the intellectual students of Voltaire and other philosophers of the enlightenment (see C. Lawrence, The art of medicine. Hippocrates, society, and utopia, *Lancet* 371 (2008), 198–199).

Places (which today would be classed as a treatise on human ecology), is among the first comprehensive studies of the effects of climate on the human condition.

Hippocrates writes about the significance of climate, water and soil conditions for the material, psychological and physical constitutions of the inhabitants of a country. He proposed a relationship between people's habits and characteristics in various places and the climatic conditions of their environment. One of his hypotheses was that fertile landscapes produced "soft" individuals, and that less fruitful areas produced "heroic" individuals. Hippocrates thought that the people who lived in what we now call Iraq and Iran were much gentler than the Greeks because they lived in a gentler climate.

For Hippocrates, nature in the form of climate was a standard and guiding principle for a diagnosis of health and illness. A natural life (centuries later this conviction found an echo in the work of the French philosopher Montesquieu (1689–1755)) meant living in harmony with oneself as formed by nature and one's own climate in particular.

The second phase began during the Enlightenment period itself, as the discussion of the impact of climate on humans and civilizations was revived with renewed intensity. Scientific academies were anxious to find the "truth". One way of doing so was by conducting competitions, looking for the best answers to such questions as this one from a 1743 competition:

> "Will the various temperaments of people also be influenced by the climate under which they were born?"

The most important commentaries and conclusions about the role of climate from this period of history have created a cultural tradition whose effects one feels even today. People adopted the principle, as Montesquieu expressed it, that there was no mightier kingdom than that of climate. For the German philosopher Georg Wilhelm Friedrich Hegel (1770–1831) it went without saying that a "culture" could actually develop only within the framework of a moderate climate. The great encyclopedias of this time imputed it as reality that ethnic differences are an expression of climatic differences.

Montesquieu argued in his theory of the division of power (first published in 1748) that there is no best form of government, but rather that institutions and justice in a state must harmonize with the given natural conditions and "nature" of the people. Montesquieu asserts that observable human variety or ethnic diversity is the result of different climatic conditions. The influence of climate on the human character became for Montesquieu the most important factor in explaining differing societal and cultural phenomena, be they political institutions, family structures or philosophical systems. According to his theory, people are cognitively and physically more active in cold climate zones than people in warm climate areas.

The German philosopher Johann Gottfried Herder (1744–1803) considered the climate problem in detail, but certainly in a more skeptical way than Montesquieu, in his primary work *Ideas about the Philosophy of the History of the Human Race*. Herder raised the basic question:

> "What is Climate and What Effect Does It Have upon the Formation of People in Body and Spirit?"

He emphasized right from the beginning of this treatise that our understanding of climate is "difficult and deceptive". But especially daring were conclusions from uncertain climatic knowledge regarding

> "whole peoples and world regions, yes, indeed even on the finest achievements of the human spirit and the random institutions of society."

Opposing historical examples always refuted conclusions such as those of Montesquieu. In spite of his reservation, Herder stressed:

> "We are malleable clay in the hand of the climate, but the climate's fingers themselves produce a great variety of forms, develop so variously and the forces that work against him are manifold. Therefore perhaps only the genius of the human race will be able to disentangle the relations among all these forces."

In spite of Herder's critical reservations, the idea that the climate has an important, determining effect on people and civilization was accorded the status of textbook knowledge in the 19th century.

Thus, the Austrian writer and geographer Friedrich Umlauft (1844–1923) popularized this very point and without provisos in his textbook of 1891: *The Atmosphere. Foundations of Meteorology and Climatology according to the Newest Research*:[25]

> "And now first consider man! ... Because Earth is not merely his living place but also the school of humanity, we must connect racial, national and cultural differences first of all with climate conditions. How different is climate in dealing with people. To some, climate bestows bountiful benefits so that people are enticed into comfortable lightheartedness. Others, forced through the hard school of unavoidable efforts and deprivations, are led to their full development of bodily and spiritual strength ... So the literature of a people is mysteriously connected with the meteorological elements of their part of the globe. The same holds for philosophical teaching systems. So the whole human culture connects with the conditions and processes of the atmosphere. Therefore ... assertions are correct that northern Europe has to thank its rain in all seasons for its position of having the world's highest culture, just as China had in earlier times a high civilization because of its summer rains."[26]

The third phase of reflections about the impact climate on society and individuals began in the second half of the 19th century and lasted until the late thirties of the last century. Participants in this international and interdisciplinary discussion were scholars from anthropology, history, medicine, geography, sociology, and so on. This discussion is marked by a decided attempt not only to postulate, but also to quantify, the influence of climate on society and people.

The American geographical researcher Ellsworth Huntington (1876–1947) of Yale University belongs among the most important authors of this time[27] who enthusiastically studied the question of the

---

[25] Umlauft, Friedrich: *Das Luftmeer. Die Greundzüge der Meteorologue und Klimatologie nach den neuesten Forschungen gemeinfasslich dargestellt*. Wien, Pest, Leipzig, Hartleben's Verlag, 1891.

[26] The description of the rain conditions was correct; see Fig. 2.

[27] See the biography by G. J. Martin, *Ellsworth Huntington. His Life and Thought* (The Shoe String Press Inc., Connecticut, 1973).

effects of climate on individuals, the economy, politics and society. He appears to have been especially successful in his time in disseminating his ideas, and is still famous, or infamous, today in certain circles.

In his main treatise, *Civilization and Climate*, first published in 1915, Huntington defends his conviction that climate must be understood as the causative factor in the history of mankind. For the "geographical distribution of human progress", as Huntington called it, the actual climatic conditions were a decisive factor for the development of civilization, along with the factors of "race" and the "level of cultural development". The rise and fall of entire civilizations and climatic conditions go hand in hand.

According to Huntington, optimal climatic conditions, that is, a certain combination of temperature and temperature variability, determine the economic performance of a society and the health of its citizens. Each deviation from the climatic optimum has in consequence a diminishment of well-being and work performance. If the climate changes or individuals move into a different climatic region, then their physical and cognitive performance, as well as their health, also change.

The substance of the work of Huntington's generation of scholars favoring *climate determinism* cannot easily be summarized, apart from their common attempt to relate a chaotic multiplicity of physical, psychological and social phenomena to the prevailing climate. The list of suggested climate variables and their alleged effects is almost arbitrary, and obviously only limited by the imagination of the thinker. It extends from conventional measurements as temperature, humidity, and windiness to exotic measurements as magnetic storms, concentration of ozone in the atmosphere, number of sunspots or phases of the moon. An enumeration of the effects includes, for example, life expectancy, crime rates, the fall of the Roman Empire, tuberculosis, stock market movements, suicide rates, intelligence, the number of workers, or marriages; but also commercial crises, the number of "serious" books checked out of public libraries, political revolutions, religious wars, police arrests, disturbances or crime rates.

Most of Huntington's quantitative deductions are based on the production statistics of pieceworkers from the years 1910 to 1913 in the New England states of the USA. Huntington linked the number of pieces produced each month by the workers with the average outside temperature.

In this way he determined the dependence of "climatic energy" on the time of year and work temperature.

He found the ideal was an outside temperature of about 15°C. The piece numbers show a slight rise in "efficiency" of the piece workers in January, thereafter a constant rise until a maximum in the month of June. In the course of the summer the numbers fall, but around the end of October/beginning of November they reach the highest point again. In a similar way the intellectual performances of students and armed forces cadets were analyzed.

Based on these claims about the link between temperatures and performance, Huntington constructed a world map of "climatic energy" (Fig. 14) using the average temperatures from approximately 1,100 weather stations around the world. The tacit premise of this procedure was that the connections found for New England would prove correct globally. On a second map, the "civilization levels" of the regions of the world are displayed. These civilization levels were determined from an inquiry among 50 scholars from 15 countries (Fig. 14). The two world maps resemble one another; for Huntington, proof that climate has and will continue to have a decided influence on the civilized and cultural development of various regions of the world.

As a result of this hypothesis, the most stimulating climate in Europe prevails in the cities in the rectangle bordered by Liverpool, Copenhagen, Berlin and Paris. One finds a range of candidates for the best climate on the North American continent, as for example, the Pacific Northwest (Seattle, Vancouver), New Hampshire in New England to New York City; but New Zealand and a portion of Australia also have a good climate.

On the basis of this insight, Huntington strongly recommended that those founding the United Nations after World War II make their headquarters in Providence, Rhode Island, because in this location was to be found the most productive climate in the world.

As indicated, the dimensions climate and race were understood as complementary but also competing factors for the "success" of civilizations. Thus it is not surprising that Huntington was also a leading member of the American eugenics movement (President of the Board of Directors of the American Eugenics Society from 1934 to 1938) in the 1920s and 1930s.

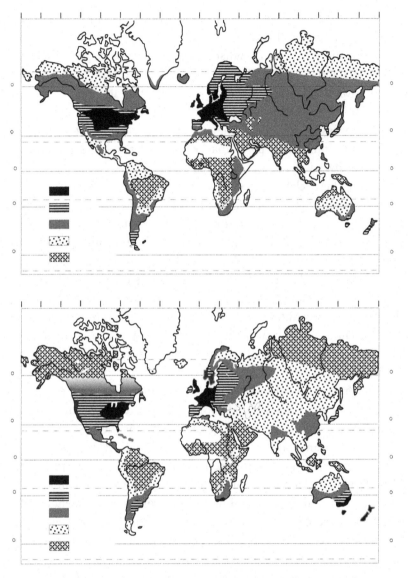

Fig. 14. Two key diagrams in Huntington's analysis. **Top:** The global distribution of people's energy as derived from the climatic conditions. **Bottom:** Global distribution of level of civilization as derived from a survey among international experts.

Eugenics as defined by Francis Galton (1822–1911)

"is the study of all agencies under human control which can improve or impair the racial quality of future generations."

Eugenics represented the contemporary scientific approach to dealing with the widespread concerns regarding a "degeneration" of mankind. Its advocates regarded it as a social philosophy for the improvement of human hereditary traits through various forms of intervention. In its program of human breeding, selection and extermination, eugenics offered a political answer to the alleged threat.

One of the theses that played an important role was that of the adaptation of different races to climatic conditions. The concept of race was one of the central categories of eugenics. For eugenicists like Huntington and his fellows, a successful adaptation to actual geographical living conditions, in particular to climate, was a central problem. Thus, the adaptation process was not seen as a historically limited necessity, or as a kind of "opportunistically" flexible approach to adapting to the environment, but rather as a universal perfecting process that was determined, at least partially, and limited by racially specific abilities or possibilities.

With these ideas in mind, certain grotesque pronouncements by Huntington become understandable, such as that African Americans had not disappeared in the northern regions of the United States only because of continual replenishment through immigration from the southern states. Similarly, Americans of Scandinavian derivation were in comparison not very successful in the dry and sunny regions of the United States. Their death rates were so high that they would have become "exterminated" in a few generations without new immigration. In the moist Pacific Northwest of the USA, on the other hand, the Scandinavians flourished immediately. The reason for success and degeneration was obvious to Huntington: under the climatic conditions of one's "homeland", which pursues one like an inborn and thus insurmountable fate, one can indeed adapt to cultural realities successfully, but one cannot adapt to a strange climate. And because, according to Huntington, climatic conditions play a decisive role, certain regions of the world are either ideal or annihilating. If the climate changes, the successfully adapted person is condemned.

People should reach the highest levels of civilization when their predispositions fit perfectly to "their" climate.

The Russian-American social scientist, Pitirim A. Sorokin (1889–1968), still a valued critic of Huntington's climate determinism, first formulated his dissenting view in 1928.[28] His critique was based on the logic pursued by Huntington himself. He did not object to attempts to conduct a quantitative investigation into the relationship between society and climate, but showed that the information Huntington used were in many cases completely inadequate and led to spurious correlations. Some of the data were extremely fragmentary, or simply in contradiction to other empirical evidence. Sorokin refers to other plausible, and presumably better, interpretations of Huntington's quantitative results. The data of other researchers at the beginning of the century indicate that there is absolutely no clear and uniform influence of climatic factors on work efficiency. Finally, the hourly and daily variations in work efficiency are much larger than the deviations that one can observe in connection with climatic factors. Huntington's chosen data collection and his methods, the selection procedures or the statistical selection for the processing of data, led to systematically false conclusions, so that Huntington's results as a whole appear questionable.

Ellsworth Huntington was unimpressed with Sorokin's criticism. He did not change his central thesis. On the contrary, he communicated his conviction about climate determination of human behavior further in subsequent and successful publications. The two maps, shown as Fig. 14, of the spatial distribution of "climatic energy" and "civilization level" also appeared in unchanged form in his last book, published in 1945 shortly before his death.

The American economist William Nordhaus (1941–), whose formulation of the Global Warming problem as an economic issue (see Sec. 4.7) influenced the worldwide political discussion, dealt in 1994 with Huntington and his hypotheses and made clear that at least the economic dimension of Huntington's analysis is flawed.[29]

---

[28] P. Sorokin, *Contemporary Sociological Theories* (Harper Row Publishers, New York, 1928).
[29] W. Nordhaus, "The ghosts of climate past and the specters of climate future", in Nakicenovic, Nordhaus, Richels, Toth (Hrsg.): *Integrative Assessment of Mitigation, Impact and Adaptation to Climate Change* (IIASA, Laxenburg, 1994), pp. S. 35–62.

The view that climate influences the way societies develop is also evident in Marxist-influenced interpretations of history. There the notion prevails that the climate and other natural environmental factors present a framework within which the dynamic of class struggle unfolds. Karl-Heinz Bernhardt (1935–), head of the Institute of Meteorology of the (then East) Berlin Humboldt University, noted:[30]

> "For the capitalistic social order Marx noted ... that the most fruitful soil is in no way the most suitable for the growth of capitalistic production methods. For the latter this 'requires the rule of mankind over nature', so that not the tropical climate but the temperate zone is the motherland of capital. 'The need to socially control and manage the natural forces, to develop techniques for large-scale use, thereby domesticating them through the works of people's hands, plays the most decisive role in the history of industry.' Climate's influence on social development — from the anthroposociogenic development of the primitive society and the developed class society, as well as the role of climate and its change as a factor of the geographical milieu relevant for the formation of the developing socialist society — obviously requires still more thorough research and discussion. This research must not only deal with geographical determinism and such unscientific, chauvinistic and reactionary assertions, such as that the result of wars would have been decided by a favorable climate, or that the inhabitants of cyclonic regions would govern the world (Huntington)."

Putting aside the concerns of method-inspired critics, the theses of Ellsworth Huntington and other climatological determinists are problematic because their radical climatological determinism hides dangers beyond questions of objectivity and scientific methodology.

The real danger of climate determinism is different: a consequence of climate determinism is that the self-determined range of human decisions becomes irrelevant, and that history is no longer the result of human actions. Independent human activity and its possibilities are replaced by a geographical determinism, and so reduced by factors over which man and society finally have no influence. People become playthings of the climate

---

[30] K. Bernhardt Klima und Gesellschaft, *Z. Meteorol* 31 (1981) 71–82.

system, subjecting themselves to the law of nature. Such an attitude demands almost unlimited, perhaps unintentional, agreement with the existing social and political order, which, supposedly for geographical reasons, must be as it is. A political regime can then claim to act in harmony with naturally given requirements and to be obliged to enforce certain rules on its citizens. Violating these rules may be described as threatening or even destroying society's natural living conditions and resources. One should not disobey the laws of climate, so it can then be argued, because of the risk that climate will avenge itself, in the truest sense of the word.

Climate determinism theories not only favored a Eurocentric reconstruction of human history, but also claimed that the future course of world history *must* necessarily be located within the limits of this framework. Since eugenicists and racists argued for the power of inborn qualities over human conduct, climate determinism assisted in developing an overall ideological framework for a strong form of racist ethnocentrism; marked differences between individuals were and will be ascribed to climate. Ethnic identity was and continues to be, in many minds, inseparably linked with climate. The popularity of this apparently fateful entanglement is seen in the fact that dominant societies saw themselves enjoying favorable climatic regions, while "barbarians" and uncivilized people naturally had to reside in climatically disadvantageous regions.

The teaching of climate determinism came to an abrupt end in the middle of the 20th century. In the period after World War II the question of the influence of climate on people and society hardly played a role any longer among social scientists. The intellectual and political proximity of climate determinism to racial theories and to National Socialist ideology silenced climate determinism in the post-war years. Climate determinism was largely ignored in scholarly discourse, in science and in the social sciences. Climate was assigned to play the role not of a cause but a constraint of human conduct.

As the German geographer Wilhelm Lauer (1923–2007) formulated it in 1981:

"Climate shapes the theatre in which human existence — the history of the human race — takes place, sets borders for that which can happen on the earth, but certainly does not determine what happens or will happen.

Climate introduces problems that man has to solve. Whether he solves them, or how he solves them, is left to his imagination, his will, and his formative activities. Or, expressed in a metaphor: Climate does not compose the text for the development drama of mankind, it does not write the movie script. That man does alone."[31]

But for the public at large, climate determinism did not lose its credibility quite as abruptly, it still appears to be a viable perspective in everyday life and in popular accounts of the relation of climate and society.

The German physician and social psychologist Willy Hellpach (1877–1955)[32] described the inhabitants of northern and southern regions respectively as follows:

"Prevalent in the North ... are the character traits of sobriety, harshness, restraint, imperturbability, readiness of exertion, patience, stamina, rigidity, and the resolute employment of reason and determination. The prevalent traits of the South are liveliness, excitability, impulsiveness, engagement with the spheres of feelings and imagination, a phlegmatic going-with-the-flow or momentary flare-ups. Within a nation, the northerners are more practical, reliable, but inaccessible, and the southerners devoted to fine arts, accessible (sociable, likable, talkative), but unreliable."

His contemporary, the German sociologist and political economist Werner Sombart (1863–1941),[33] was certain, until his late creative phase, that

"soil and climate jointly decide not only the natural fertility of a country, they determine to a great extent whether the nature of a people tends to either indolence or to activity."

---

[31] W. Lauer, Klimawandel and Menschheitsgeschichte auf dem mexikanischen Hochland. Akademie der Wissenschaften und Literatur Mainz. *Abhandlungen der mathematisch-naturwissenschaftlichen Klasse*, Nr. 2, 1981.

[32] W. H. Hellpach (Kultur and Klima, 1938). In H. Wolterek (ed.), *Klima–Wetter–Mensch*, Leipzig, Quelle & Meyer, pp. 417–438 (republished in 1952). See also Nico Stehr, The ubiquity of nature: Climate and culture, *Journal of the History of the Behavioral Sciences* 32 (1996) 151–159.

[33] W. Sombart, *Vom Menschen: Versuch einer geisteswissenschaftlichen Anthropologie* (Berlin Buchholz & Weisswange, 1938).

The British meteorological journal *Weather* published in 1993 a remarkable contemporary example of climate determinism. There a Mr. Beck wrote with considerable conviction:

> "Several writers have remarked on apparent correlations between the character of the people of a region and the climate prevailing there ... intolerant acts have often been committed by people from areas in mid-latitudes where seasonal temperature extremes are large, as in areas with continental climate. In the 1930s, fascism took over in Spain, Germany and Austria; all are continental countries with annual temperature differences (TD) generally averaging about 20°C (except for Southern Italy, with a TD of only 15°C, but early support for the fascist Mussolini was said to be weak there) ... Many of the states of the USA which retain capital punishment have TD values of over 20°C, which is high compared with most other 'western' nations ... It may never be possible to prove absolutely that a mild climate in mid-latitudes helps to foster a tolerant society or that an extreme climate may predispose people towards intolerance. However, the historical record is highly suggestive and if this is recognized it could help to identify potential problem areas in the field of human relations so that timely action can be taken to mitigate threats to peace..."[34]

The causally responsible mechanism should be the absence of extreme seasonal climate differences, as Beck emphasizes:

> "... a relaxed attitude because there is no need to make elaborate plans to cope with the rigors of a cold winter and/or a very hot summer. However, where TD is large, the pace of life is driven by the seasons, enforcing the discipline of timely preparation for the extremes; here, less relaxed mental attitudes may develop."

As one may easily see, these pronouncements differentiate themselves only minimally from the assertions made 100 years earlier by Friedrich Umlauft.

---

[34] R. A. Beck, Viewpoint: Climate, liberalism and intolerance, *Weather* 48 (1993) 63–64.

It would certainly be an attractive assignment for a social scientist to investigate empirically to what extent ideas of climatic determinism are part of everyday thinking, and to what extent this is linked with other determinisms, especially genetic determinism (and racism).

Recently there have been occasional attempts to deal again with the classical questioning of the significance of climate for people — for example, in the climate impact research on the consequences of climate — mostly without being aware, however, that there is a long tradition of analyzing the consequences of climate for people and society.

4

Climate as Risk and Hazard

In this chapter we want to deal with the climate as a variable, at times even capricious constraint on human existence. Unlike the perspective of climate determinism in the preceding chapter, we deal here with erratic and uneven characteristics of the climate, so that society is confronted by an unreliable existential resource; and therefore one can justifiably speak of the influence of climate on society representing a risk and hazard. Although climate appears essentially as a constant for the individual, infrequently interrupted by extreme events and perhaps by some moister or warmer years, the analysis of a long series of observations shows that significant climatic variations do occur. We begin this chapter with a series of examples displaying climate variability.

A historical example is provided by Fig. 15 — it shows the analysis by Eduard Brückner[35] of 5-yearly averages of rainfall in the 18th century and wheat prices in England. There are two things to learn from this diagram.

---

[35] We will refer to Eduard Brückner's work again in this chapter. An anthology containing many of his articles translated into English has been edited by Nico Stehr and Hans von Storch, 2000: *Eduard Brückner — The Sources and Consequences of Climate Change and Climate Variability in Historical Times*. Kluwer Academic Publisher ISBN 0-7923-6128-8, 338 pp. Figure 15 is from the presentation "Climate Variability and Mass Migration," given at the Kaiserliche Akademie der Wissenschaften, Vienna, 1912; it is reprinted in English in the anthology on pp. 285–297.

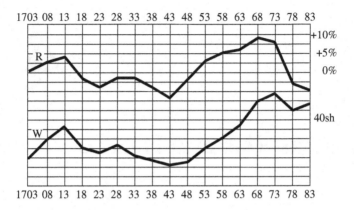

Fig. 15. 5-yearly averages of rainfall amounts (R) and wheat prices (W) in England in the 18th century according to Brückner's analysis. The vertical axis is divided in increments of 2.5% for rainfall and 2 shillings per Imperial Quarter. Redrawn from original.

First, that precipitation is changing from a 5-year segment to 5-year segment; variations are of the order of ±5%. The deviations from the long-term mean extend across many years, which was at the time of Brückner's publication in 1912 a significant observation, as it was thought that averaging across, say, 30 years, would largely eliminate all variations.

The second issue is the price of wheat, which Brückner claimed was closely related to the rainfall amount, namely that in the maritime climate of England an increase of rainfall would be associated with smaller harvests and thus higher prices. While this may have been the case in earlier times, when international commerce was less important, Brückner pointed out that at the end of the record, the link would break down, and he related this breakdown to other factors, in particular, political factors.

The presence of climate variability on time scales of hundreds of years is also demonstrated, for example, by the choice of the name "Greenland" by the Vikings in the Medieval Warm Period. From the 11th to 13th centuries, Greenland was still "green" due to the mild climate. Today hardly anyone would choose such a name.[36]

---

[36] Even though one cannot completely reject the possibility that this naming was an advertisement to attract immigrants, similar to the suppression of reports about tornadoes as illustrated in Fig. 5!

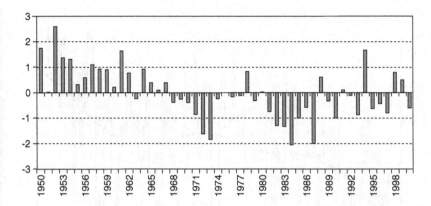

Fig. 16. Index for rainfall amounts in the country of Niger, 1950–2000.
Source: Direction de la Météorologie Nationale, Niger. Redrawn from original.

The third example deals with rainfall in the African country of Niger (which was affected by the so-called Sahel drought) that brought adverse conditions to agriculture in large swaths of Africa for many years (see Fig. 16). Hypotheses have been discussed to relate this drought to sea surface temperature conditions.

Today, society's concern is concentrated on a possible anthropogenic change in climate due to the emission of greenhouse gases into the earth's atmosphere. Such links are also claimed in the case of the Sahel drought, but the jury is still out …

The last example shows temperature development in the country of Denmark (Fig. 17). Obviously, the climate has become significantly warmer in this country, by about 1.5°C. From the beginning of the record in 1873, the temperature rose continuously and steadily by 0.7°C until about 1950; then came a period of about 20 years with a small decrease, until about 1980 when the warming returned at an unprecedented gradient. Within a mere 25 years, the temperature has risen by another 0.7°C, until now. This increase is fully consistent with the explanation of anthropogenic warming, which is further supported by the continuation of the warming. Of recent years, even the coldest records were warmer than the warmest at the beginning of the record. It seems that in this case, the jury is no longer out …

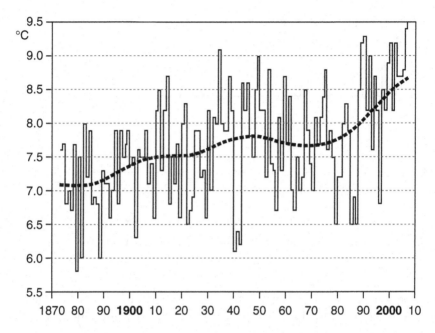

Fig. 17. Development of annual mean air temperature in Denmark derived from homogeneous temperature records collected throughout the country.

Source: Danmarks Meteorologiske Institut. Redrawn from original.

In the following sections, we will discuss both the scientific construction of climate change and the social construction: which of the two constructions will decisively influence the formation of climate perception and climate politics remains unclear. Certainly, the two constructions are not clear cut dichotomies; the social construction of climate is present in scientific discourse, which is certainly not helping science to pursue what many see at is ultimate goal, namely of "objectivity".

In Sec. 4.1, we recapitulate the historical scientific discussion of the theme of "climate change", a discussion that in many respects resembles the current discussion of anthropogenic climate change. Then we turn to the scientific concept of climate changes in Secs. 4.2 and 4.4. As an excursion, we add an account of the military dimension of climate and weather manipulations in Sec. 4.3. In Sec. 4.5 we discuss lay knowledge about climate change, and review a number of historical cases in Sec. 4.6.

Finally, we discuss in Sec. 4.7 the effect of both constructions of climate change on society and politics.

4.1. The History of Ideas about Climate Change

The question of climate changes and their causes in geological or historic time was always of interest for climate research, but it was certainly not among the dominant research questions. At the beginning of the "scientification" of climate research, in the middle of the 19th century, climate change was largely ignored. Only at the end of the 19th century did climate variability in historical time — above all due to observations and analysis by Eduard Brückner (1862–1927), see Fig. 18 — become an important

Fig. 18. Eduard Brückner, together with his teacher and colleague Albrecht Penck, was instrumental in detecting traces of previous ice ages in the European Alps.

research topic. However, it soon retreated into the background again. Only in our time has the problem of possible climate changes and their causes again become one of the central objects of climate research.

We are confronted today by the fact that the intensive and controversial discussion about global climate change caused by people is taking place not only among scientists, but also among the public. In this connection, the "Greenhouse Effect" concept is today generally common knowledge. Scientists appear alarmed, and some address the public directly, warning of an imminent climate catastrophe. One gets the impression that a completely new environmental threat has emerged. In actual fact, this is not the case.

A similar discussion among scientists took place a century ago, when it became clear to a number of climatologists that our climate changes not only in geological time, but rather also within centuries and even decades. This observation was supported by data on the water level of lakes without outlets, as for example, the Caspian Sea. It was asked whether the altered water level was a result of human activities, or brought about by a natural climate variation. Large-scale deforestation and the cultivation of large stretches of land were suspected to be reasons for anthropogenic climate change. Scientists were partly convinced that these changes brought about positive developments (in the sense of "rain follows the plow"). But more often, people referred to negative consequences of climate change.

It is especially noteworthy that these discussions were not just limited to science. Some contemporary scientists appealed directly to the public, and demanded measures that one would call today climate policies or climate protection. They advocated the prevention of further climate changes with negative commercial, social and political effects. Other scientists were convinced that the observed changes would be a natural phenomenon, possibly associated with some cosmic processes, to which society must "adjust". In some European states, parliamentary and governmental commissions were formed to discuss steps regarding the problem.

In the following section, we want to discuss two of the prominent protagonists of climate discussion a century ago, namely the already mentioned Viennese professors Eduard Brückner and Julius von Hann. For a long time they held chairs in geography and meteorology, respectively, at the University of Vienna.[37] They represented opposing positions in the question of the meaning of historical climate change.

In 1890, Brückner published his principal work, *Climate Changes since 1700*. From the variations in the water level of the largest inland sea in the world, the Caspian Sea, he decided that the observed changes must have a climatic, 35-year quasi-periodic cause. The highest water levels are the result of cooler and wetter atmospheric conditions, while the lowest will be produced by dry, warm weather. In another paper, Brückner analyzed the connection between rainfall and related water levels on a global scale. He found that the hundred-year variations take place in all regions of the world. Brückner emphasized that the cause of the quasi-periodicity he established was unclear.

Brückner was very interested in the economic, social and political consequences of climate changes. He dealt with the question of the influence of climate changes on migration, crop yields and balances of trade, as well as on health and changes in the international political power balance. He began with the idea that variations in rainfall amounts have a direct effect on agricultural production. Then he found that in western and middle Europe, above-average agricultural harvests occurred in warm and dry weather periods (Fig. 15). Conversely, a comparable decline in productivity took place in wet and cool atmospheric periods. For both continental Russia and the central United States, he found the inverse effect. This geographical pattern of climate changes affected the migration from Europe to the United States (Fig. 19). Worsened agricultural conditions in Europe's maritime climate parallel improved condition in the continental North American climate, and vice versa. Brückner found this point of view confirmed through emigration numbers and precipitation statistics (Fig. 19).

Brückner presented his research results in both oral and written form. In lectures and newspaper articles, he addressed the public at large, as well as professional groups that were especially affected by climate changes, for example, farmers. His ideas were discussed in the contemporary press.

---

[37] For an account of their work and ideas, refer to the anthology by Stehr and von Storch, *op cit.*, in particular the Introduction, pp. 1–24.

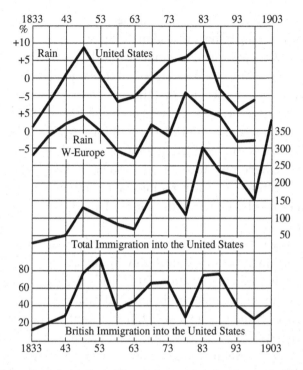

Fig. 19. Brückner's analysis of rainfall variations and emigration to the United States in the 19th century.

Julius von Hann had a different approach to climatology. He was considered one of the most important meteorologists of his generation and was the author of the first textbook on climatology, first published in 1883. For him the representation of the climatic state on the basis of reliable observations was most important. With this representation, a basis for both the understanding of different meteorological phenomena and climatological consulting should be possible. Von Hann avoided discussion of the influence of climate on society.

In his textbook, he dealt only marginally with the concept of climate change. The discussion of climate at that time was primarily concerned with periodicity. In this sense, von Hann distinguished between "progressive" changes (that is, lasting changes or, as we would say today, "climate change") and "cyclical" changes (that is, fluctuations and oscillations around a mean state) with certain characteristic periods. From the recorded observational data at his disposal, von Hann deduced that there is no convincing proof in historic periods for "progressive" climate change in different continents and countries. He saw just as little proof in his observational data of a reaction of the climate system to the periodic variation in the number of sunspots. Therefore, he rejected the notion of any effect of sunspot variations on earth's climate. However, he was much more open and positive about Brückner's 35-year periodicity of climate variation, derived from a large number of empirical observations with opposite tendencies in continental and maritime climates, than about speculations on the effect of sunspots. Also, contradictory findings so far about climate changes in different geographical areas and times could be explained by the fact that these reflected different sections of the 35-year cycle.

In many respects the situation at the end of the preceding century was comparable with today's — scientists found it increasingly evident that climate is not constant, but rather undergoes significant changes within centuries and decades. At the same time, it was clear that climate can change both systematically, as a reaction to human behavior (in von Hann's terminology "progressive"), and temporarily (in von Hann's words "cyclical"), due to natural processes. The causes for natural climate variability were unknown. Speculative hypotheses involved solar activity or other "cosmic" factors. Comparable to some contemporary reactions, a number of scientists prematurely and erroneously interpreted relatively short segments of slow natural climate changes as indicators of systematic alterations. For instance, some scientists interpreted Brückner's quasi-periodic variations as permanent consequences of deforestation and other modifications of land use.

Considering the fact that climate has a considerable influence on certain sectors of the economy and social institutions, the scientists saw, then as now, that the question was whether they should merely inform the public, or instead explicitly warn the public of the expected climate changes. Some scientists, for example, von Hann, decided to keep to accurate measurements and analyses of observational data and to communicate exclusively with other scientists. On the contrary, others, like Brückner, felt ethically obliged to turn directly to the public. In contrast particularly to environmentally conscious "advocate" scientists of today, Brückner did not demand any specific political measures for abating climate change. Other scientists, however, did not hesitate. At the end of the 19th century, the American F. B. Hough called for comprehensive reforestation measures in North America in the name of the American Association for the Advancement of Science (AAAS) in order to avoid further drying of the North American continent. The advocates of this thesis of anthropogenic climate changes in the last century actually had a certain influence on public administration and politics. In several states, governmental and parliamentary commissions were instituted to deal with the implications of climate change.

The intense scientific and public discussion of climate variability at the end of the 19th and beginning of the 20th century quickly disappeared from the scientific and public agenda. A new consensus, dominating until the 1970s, mostly removed climate change from scientific consideration. Thus it was maintained that climate variations were only episodic in nature and, because of the inherent climatic equilibrium, small and insignificant with respect to impact.

This section about the history of science has shown that the growing popular scientific genre of climate research, especially the public explanation of the climate problem, is not new. Today, just as a hundred years ago, scientists with differing backgrounds and perceptions are engaged in these discussions. Also, it is not only today that the uncertainties and doubts linked to the data in the projection of climate developments draw attention — von Hann had this critical attitude one hundred years ago.

Today, many observers feel that the global perspective represents entirely novel kinds of views and problems. This is not accurate. As our case demonstrates, already at end-19th century, global climate variability and change was detected, and even predicted for a few decades into the future. Eduard Brückner was convinced that our climate is a global system, experiencing variations on the global scale and on time scales of decades and centuries.

But why have the once intense and sometimes passionate discussions of climate variations and their social effects at the turn of the 19th and 20th centuries become almost completely silent and fallen into oblivion? Certainly there have been other important problems — the First World

War, serious economic crises, and the ensuing totalitarian regimes, which doubtlessly suppressed interest in the questions of nature's variable effect on society and society on nature.

4.2. Natural Climate Variability

Climate variations related to natural processes can be internal activities in the climate system, or can be caused by external factors, such as changes in the solar output, or in the presence of stratospheric aerosols due to volcanic eruptions. When we speak of "variations", we refer to deviations from a "normal condition". These deviations persist for different lengths of time, and will alternate with variations in the opposite direction. After a cool phase follows a warm phase, after a dry one a wet one, and so on. Thus, the "normal condition" is an imaginary value, because there is no "normal" in the geological history of Earth. The positive and negative deviations do not equalize in the long run. "Normal" and "deviations", or as they are sometimes called "anomalies", are mathematical constructions. The "World Meteorological Organization" has set the averaging interval to be 30 years.[38] This length of 30 years is not a natural constant, but a socially agreed convention that matches with the time horizon of human experience.

One must not perceive the sequence of variations as a periodic chronology, such as the annual march of temperature or the tidal water level variations. One can predict periodic processes ad infinitum: the northern summer will be warmer than its northern winter in the fourth millennium, as now. However, one cannot predict climate changes in this manner. Today the predictions for variations in the average temperature of the coming season have little accuracy, even if they are limited to simple statements like "warmer than normal" or "colder than normal". The sequence of variations is irregular; after ten warm years there can just as well be three cold or three warm years; after many years of decreased precipitation, as in the case of the sub-Saharan Sahel in the 1970s and 1980, there can be a period of alternating moist and dry years (cf. Fig. 16).

---

[38] Rumours have it that the specific choice of 30 years is related to the quasi-period of 30–35-year cycles proposed by Brückner.

Hann's notions of "cyclical" and "progressive" climate changes do not fit well.

"Progressive" changes are irreversible variations that can be caused by systematic changes in radiation budget or in the surface condition. Pertaining to this category are the anthropogenic changes that we will discuss in the next section. The changes in the positions of the continents, or the formation of mountains, which take place on very long time scales of millions of years, are also "progressive" changes.

The word "cyclical" implies not only the temporary character of the changes, but also a periodicity. The trick was to consider "non-progressive" variations as being composed of a finite number of "waves" with characteristic periods. In order to be able to make a prediction, one needed only to determine, by the method of "harmonic analysis", these periodicities and the condition of the preceding days or years, depending on the kind of prediction problem. However, as already mentioned, this approach was, and is, fundamentally flawed.

In the renowned *Manual of Meteorology* of 1936, Sir Napier Shaw (1854–1945) presented several pages showing supposed periodicities in the climate system, from a few days to months. This list included the results of many studies done in the previous decades. Harmonic analysis was employed not only in meteorology but also in many other areas, from economic cycles to earthquake prediction. The enthusiasm with which harmonic analysis was accepted was also reflected in the founding of a *Society of Cycles*. One of its founding members was Ellsworth Huntington. Even today, occasional attempts are made to filter "significant periodicities" out of data and to use them for predictive purposes.

The fact that practically every conceivable period could be found in meteorological data sets should have made scientists suspicious early on. The regular failure of predictions should also have been a warning. But only the analysis of the Russian mathematical statistician and economist E. Slutsky (1880–1948) eventually ended the hambug in 1937.[39] He demonstrated that one finds periodicities in any finite time series, even if

---

[39] E. Slutsky, The summation of random causes as the source of cyclic processes, *Econometrica* 5 (1937) 105–146. The paper was originally published in Russian in 1927.

by construction there are no such regular variations in the data. Entirely randomly composed variations may be artificially separated into periodic components in this manner.

By now, Slutsky's finding has become an integral part of the sound methodological arsenal in meteorology, oceanography and related disciplines. But in certain backward quarters, the old naivety still prevails; sometimes it takes time to overcome convenient errors, in particular when the new findings represent an obstacle in supporting convenient and simple claims.

Today we conceive climatic time series as a mixture and sum of externally caused, "deterministic" components (such as volcanic eruptions) and internally produced non-periodic variations. The changes in the parameter of the earth's orbit (the shape of the earth's orbit around the sun and the tilt of the earth in this orbit) within tens of thousands of years have a periodic effect on the earth's climate. This Milanković theory (named after the Serbian Astrophysicist Milutin Milanković) is powerful to explain many, but not all, aspects of the alternations of glacial and interglacial periods.

The possibility that sunspot cycles have a periodic influence on the earth's climate was categorically denied after many futile attempts at explanation, and was discussed again for the first time in many years only after two climatologists from Boulder and Berlin, Harry van Loon and Karin Labitzke, discovered intriguing new phenomena in this area. Now, the jury is out again... and time plus new data will help to eventually sort out the role of the sun.

The shortest period of climate variability spans only a few days. Such "weather variations" appear in the form of passing storms or lasting high-pressure situations ("blocking") in the extratropics. For meteorologists, these disturbances represent instabilities and nonlinearities in the large-scale turbulent stream flow of the extratropical atmospheric circulation (cf. Fig. 11).

Their frequency and intensity are variable and can be understood, in a good approximation, as randomly distributed. Such disturbances are also responsible for the occurrence of extreme storm events. Because of the statistical character of the frequency and strength of storms, extreme events like a "100-year storm" or a "1,000-year flooding" are possible any time. For any given time and location, the probability of such an event is minimal, but not null.

Until now, climate research produced few results that concerned the causes of variations from year to year. An exception is the tropical "El Niño" phenomenon. This is an irregular event that accompanies the expansion or recession of warm water in the equatorial Pacific. The results are marked precipitation anomalies; some areas in South America experience excessive rainfalls, others in Australia, droughts. Apart from the seasonal variations, there is no other natural phenomenon lasting persistently for many months besides "El Niño". These anomalies, predictable with the help of models, usually last about one year, and there is some tendency to alternation in the consecutive years. "El Niño" is hardly predictable for more than one year in advance.

For the largest part of the extratropical regions and particularly for Eurasia, "El Niño" has no significance. One assumes today that the "El Niño" phenomenon is excited by the favorable combination of a wave propagation process in the equatorial Pacific Ocean and an intensified release of thermal energy in tropical convection processes.

Moreover, on time scales of 10 to a hundred or more years, the climate system shows distinct variations. Because of missing or inhomogeneous observation data, these variations are insufficiently documented and understood. The "Little Ice Age" in northern Europe from about 1500 to 1750 is an example of a climate anomaly lasting for centuries. The "Younger Dryas" event of about 11,000 years ago, during which there was a sudden return to cold conditions in northern Europe, likewise belongs in this class of variations. The fact that climate variation also occurs on still longer time scales is demonstrated by the famous diagram displaying temperatures and CO_2 concentrations derived from an ice core covering the last 160,000 years. This ice core was extracted at the Russian Antarctic station Vostok and jointly interpreted by Russian and French scientists. The relatively warm conditions in the last 10,000 years, the sequence of ice ages in the preceding 90,000 years and the last interglacial ("Eemian") about 120,000 years ago can all be clearly seen. One also recognizes the parallel development of temperature and CO_2 concentrations — warm temperatures accompany elevated CO_2 concentrations and vice versa. Nevertheless, it is unclear whether the changed temperatures are the reason for the altered CO_2 concentrations or the other way around, or whether perhaps both of them are controlled by a third, unknown process.

Our knowledge of the natural causes of climate change is imperfect. As already mentioned in Sec. 3.2, it is disputed to what extent changes in solar activity affect the climate. This is especially valid regarding sunspots, which historically have always played an important role in considerations about the causes of climate variations.

Fundamentally long-lasting climate variations can be produced through three different processes:

1. Through external influences. The theory of the Serbian astronomer Milutin Milanković, whereby the ice ages could be explained by periodic variations in the earth's orbit, belongs in this class. We know today that these cycles can explain only a part of the succession of Ice Ages and warm ages. Other external factors concern extraterrestrial processes, especially solar activity or changes in the earth's topography. Examples are the striking rise in solar activity in geological history (more about this in Sec. 4.4) and the continental drift.

    Until the 1950s, external influences were the only way of explaining climate variations.[40]

2. Internal, "deterministic" dynamics, such as those due to nonlinear interactions within the system, have the power to produce very interesting temporal and spatial patterns. The already mentioned "Younger Dryas" event possibly belongs in this category. Here we should mention particularly "Chaos Theory", the sensational discovery by the American meteorologist Edward Lorenz in the middle of the 1960s. But hypotheses for explaining climate variability, and possibly changes, making use of this "Chaos Theory" have not been really persuasive to this point.

3. A straightforward explanation comes from the Hamburg climatologist Klaus Hasselmann (1931–), whereby physical systems can be stimulated to exhibit slow variations by exposing them to quickly changing

---

[40] This mode of thinking was presented, for instance, in E. Huntington and S. S. Visher, *Climatic Changes* (Yale University Press, New Haven, 1922).

random events, so-called "noise".[41] In climate, weather events play the role of the noise. Hasselmann's concept is found to be a valid description of many significant aspects of climate variability. It is consistent with the absence of periodic cycles and the overall stationary statistics of climate variability.

There are various procedures for studying climate variability: one approach is to analyze observation data, another to design and run detailed, realistic models of the climate system. Experiments in the strict sense are impossible, because there is only one climate system. This system is "open", i.e., exposed to a series of uncontrollable, external influences. Moreover, it is difficult to define a "margin" or a border of the climate system. What belongs or does not belong to that system? The sun does not belong to it, but the earth's atmosphere does. But what about vegetation, what about people?

The analysis of observation data suffers from the fact that these data often represent only the conditions in a rather small area for a relatively short time, so that it is difficult to reliably assess large-scale and long-term conditions. There have been global observations for only about 150 years, and this supposedly global set of data has large spatial gaps. Large areas of the Pacific and the Southern Ocean were for many years hardly traversed by ships, so there is hardly any data for these regions. Good sets of data with greater spatial resolution and quality-assured observations have existed for perhaps 30 years, since satellites were routinely employed. This data is, of course, inadequate for a description of climate variations extending for decades.

Besides instrumental data routinely collected over about the past 150 years by meteorological and oceanographic services, there is also indirect data, such as the previously mentioned ice cores. The width of tree rings, or the deposit characteristics of deep-sea sediments, also yield information about past climate variations. Extracting knowledge from "annual tree ring width" or "isotope relationships in limestone shells in ocean sediment" is in no way trivial. The result is of only limited accuracy and precision. Interpreted by experts, these pieces of information provide

---

[41] K. Hasselmann, Stochastic climate models, Part I Theory, *Tellus* **28** (1976) 473–485.

a wealth of information about climate variations in the course of hundreds, thousands, even millions of years.[42]

Climate models are complex mathematical realizations of the functions and interdependence of the components of the climate system.[43] They are approximations of the actual climate system. The atmospheric and oceanic components of climate system are best described by such models, because a significant part of their dynamics, namely hydrodynamics,[44] is at least in principle completely known.

Because the equations cannot be resolved exactly, scientists use the method of mathematical approximation, i.e., they limit their description to a "significant part" of the complex system. Because of the non-linearity of the system, this restriction introduces errors, though ideally small ones.

Thus, while hydrodynamics are satisfactorily but not perfectly represented, modeling of thermodynamic climate processes (for example, phase transitions, such as water condensation or mixing processes) is less well solved. These processes are, in the main, well understood if one operates on small, or the smallest, spatial scales. But in climate models, the smallest resolved spatial scales are many orders of magnitude larger than the microphysical scales of these thermodynamic processes. An example: the process of absorbing and reflecting radiation in the atmosphere is decisive for the formation of the climate. In this regard, clouds are of particular importance, and the process of absorption and reflection depends on the size of water drops in the clouds. For the climate model itself, the size of the drops is of no interest, but the effect on the radiation transfer is really significant. Such processes are "parametrized"; that is, the net effect on the large-scale condition variables is "estimated". The parametrization is constructed so that the simple physical principles (such as mass conservation) are fulfilled, and so that they are consistent with instrumental observation data. But above all, the parametrizations are designed so that

---

[42] Cf. van Andel, *op cit.*, or T. J. Crowley and G. R. North, *Paleoclimatology* (Oxford University Press, New York, 1991).

[43] E.g., Müller and von Storch, *op cit.*

[44] Hydrodynamics describes the flow of fluids and gases governed by the laws of conservation of mass, energy and momentum.

the climate model are capable to simulate the global climate. In this sense, all parametrizations are optimized so that they reproduce today's climate. It is then hoped that the parametrization is also valid in a somewhat altered climate, since the anticipated climate changes are in absolute physical terms, small.

Climate models can be tested only to a limited extent with regard to their ability to represent climate variability. One such test is whether or not the climate models correctly reproduce the annual cycle.

The predictability of weather, and that of the "El Nino" phenomenon, are further indicators of the model's believability. But to what extent modern models realistically simulate weather with long-term natural variability, on time scales of decades and longer periods, is at the moment still subject to intense research.

Climate models play a prominent role in climate research not only because of their ability to generate "scenarios" of future climate developments, but also above all because of their ability to make possible a "virtual, manipulable reality", within which specifically designed experiments can be conducted. As opposed to reality, climate models are closed systems. They can, at least in principle, be run as often and as long as needed, so that, as in classical physics, several statistically equivalent realizations can be generated. With such climate models, attempts can be made to understand the role of cirrus clouds in the formation of the climate system, or the influence of precipitation related to storms in the North Atlantic on the thermohaline ocean circulation.

Like the natural climate system then, climate models generate variability on all time scales, without any variation in external factors, such as solar radiation. This variability is free from periodicity — apart from daily and annual cycles — and can be well described as random. Therefore it can be predicted only for short intervals in the world of the model, as in reality. Although the spatial-temporal characteristics of this variability cannot be strictly verified by means of observation, a general consistency with empirical data can be established, and the dynamic character of the variability can be analyzed in the framework of the simulated data. In this way, insights into the stability of the Gulf Stream, or into the nature of the North Atlantic Oscillation, and so on, are obtained.

But the success of the climate model in reproducing details depends on their spatial size. In accordance with the spatial cascading discussed above, the models are more successful the larger the spatial dimension is. On the other hand, the reproduction of small spatial details is frequently unsuccessful.

4.3. Excursus: Who Owns the Weather in 2025?

A few years ago, when the authors gave a joint paper on perceived man-made climate changes, a listener approached us and told us about research carried out by the U.S. Air Force. The findings were reported back in 1996 but have since been generally ignored in the public debate; the web-page has been disabled by now. The issue was simply: He who controls the weather, controls the world.

In 1977, the UN General Assembly adopted a resolution prohibiting the hostile use of environmental modification techniques. The resulting convention (ENMOD, Convention on the Prohibition of Military or Any Other Hostile Use of Environmental Modification Technique) committed the signatories, which included the United States and the Soviet Union, to refrain from any military or other hostile use of weather modification that could result in widespread, long-lasting or severe effects on the economy and society.[45]

Thinking about the possibility to change the weather for tactical purposes has always been popular in military circles (Fig. 20). In the 1996 study, seven military officers considered again how the weather might be used as a weapon. Their task was to ensure that the United States remained the dominant power in aviation and space travel in the year 2025. The study concludes that America's airborne military forces can "own" and control the weather. This would promote the development of new technology, and that technology will provide the "warriors of the future" with undreamt-of resources for controlling the course of military conflicts, the study concluded.

The byword here is "weather modification", in the sense of increasing or decreasing the intensity of natural phenomena. Taken to an extreme, this

---

[45] See L. Ponte, *The Cooling* (Prentice-Hall Inc. Englewood Cliffs., New York, 1976).

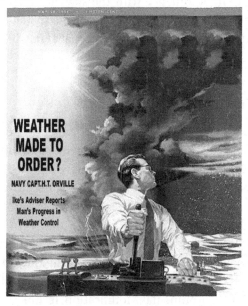

Fig. 20. Concepts of modifying the weather not only for civilian but also for military purposes caught the public's imagination in the 1950s.

could include creating entirely new weather phenomena (made-to-order weather) and manipulation of the global climate.

But because of the probable conflict with the ENMOD convention, the study concentrates on influencing weather processes in geographical areas only up to 2,000 square kilometers (800 square miles). What is at stake becomes clear when the authors of the research paper talk of a dilemma akin to that once faced by the pioneers of nuclear research. They stress that only those who are prepared to surrender strategically could want to renounce the military use of weather modification.

Specific operations to curtail an opponent's operating ability and improve one's own could include manipulation of precipitation, storms and fog, but could also involve controlling the ionosphere to guarantee dominance of worldwide communication. The research paper did not yet speak of controlling temperature.

The report explains how military encounters could be decided through weather manipulation. It cites the following example: it is the year 2025, and a South American drug cartel has purchased hundreds of Russian and Chinese fighter planes. So far, the drug barons have been able to protect their production facilities from every attack. The cartel controls the skies and is able to launch ten of its own planes for each American fighter. It also has a sophisticated French air defense system. Despite all this, the American military want to engage the enemy.

The air force meteorologists are to play a crucial role. They point out to the air force that there is a thunderstorm nearly every afternoon in the equatorial regions of South America. The U.S. Secret Service knows that the cartel pilots are reluctant to fly in or near thunderstorms. So the weather task force will not only forecast storm paths, but also to trigger or intensify thunderstorms over critical target areas. And as U.S. fighters fly in any type of weather, they are able to snatch control of the skies from the enemy. Moreover, it is likely the air force will routinely use unmanned drones to manipulate the weather by 2025.

These operations will be supported by highly developed, sophisticated technologies for data collection, weather forecasting and weather manipulation. The unmanned craft can spread cirrus clouds over areas of military deployment. Not only does this deprive people on the ground of a clear view, it also prevents them from using their infrared equipment properly. While microwave heaters create local zones of destructive interference to restrict the use of radar-controlled equipment, the naturally occurring thunderstorm is artificially intensified. It is all part of the game plan.

It is therefore possible that such systematic weather modification will become a potent, accurate and globally available weapon of war. It could be used in all conceivable conflicts. Weather is not only everywhere; it is at the same time the most implacable enemy of the ruled and of the rulers, as this report illustrates. In future, the weather may be party to a conflict.

Systematic attempts to influence the weather by technical means have existed for a considerable time. However, these efforts (for example, attempts, to control precipitation in arid areas or during droughts) have not been particularly successful so far. Rainmaking is certainly possible in certain situations. But these situations are rare and not easy to control, given the complexity of weather systems. The authors of the U.S. Air Force research paper were clearly aware of these facts. Thus, they speak of significant and rapid progress in our understanding of the variables that affect weather. They are certain that by 2025, it should be possible to identify and parametrize all important weather factors. The authors also say there must be quick and meaningful technical progress, so that micro-meteorology can develop into a discipline that is technically sound and practical. As things now stand, implementing the report's ideas appears utopian and expensive.

The major significance of the weather for the living conditions of a growing world population could, however, also cause appropriate resources to be provided for research into improving our knowledge of the weather. Summing up: it may be that by 2025, man will have taken the step from scenario planning to effective weather modification. The military uses of this knowledge are obvious.

4.4. Anthropogenic Climate Change

The global climate is largely determined by radiation budget, and by the condition that on average, incoming radiation equals the outgoing radiation. The notion of this balance was introduced in Sec. 3.2; in Fig. 21 it is shown more formally, in order to show how solar equilibrium can be altered by natural occurrences and human activity.

To begin with, the energy E, emitted by the sun, reaches the atmosphere's upper levels. This energy today amounts to 342 watts per square meter (W/m^2) and takes the form of short-wave (solar) radiation. A portion of this energy E will be reflected to space as short-wave radiation. This reflection takes place at clouds, ice surfaces, the earth's surface, etc. The spatial average of the reflected portion, the albedo, amounts to $\alpha = 0.30$. The remaining 70% of radiation reaching the earth surface will mostly be absorbed by the various surfaces of the earth, principally land and ocean, leading to a rise in temperature there. On the other hand, according to the Stefan–Boltzmann law, these surfaces emit long-wave (thermal) radiation A

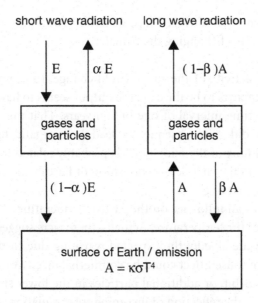

Fig. 21. Sketch of the Greenhouse Effect.

proportional to the fourth power of temperature. The higher the surface temperature is, therefore, the more energy-rich the long-wave radiation. This long-wave radiation encounters gases in the atmosphere on its way into space — gases that absorb some of the long-wave radiation and re-emit it in all directions. Thus, part of the energy returns to the surface, from which it is radiated anew. We designate this part βA. In the end the temperature at the surface attains a value so that the associated long-wave radiation $(1 - \beta)$. A reaching space matches the arriving, non-reflected short-wave radiation $(1 - \alpha)E$. In the notation in the sketch, this condition reads: $(1 - \beta)A = (1 - \alpha)E$. This is the Greenhouse Effect. Actually, the transfer of energy through the atmosphere is more complicated because, among other things, the surface loses energy through vertical heat and moisture transport. But this does not change the principal picture of the Greenhouse Effect.

If there were no Earth atmosphere, one would expect α to be much reduced (no clouds), $\beta = 0$ (no gases and aerosols); so that the global average temperature would be about −10°C. With an atmosphere present, one calculates a realistic value of +15°C.

From this somewhat abridged representation comes a series of four possible mechanisms of changing the climate:

1) The radiative activity of the sun E could change. In geological time, this actually appears to be the case; the output seems to have risen distinctly over a time interval of one billion years. That this increase in solar activity did not consequently lead to dramatic temperature changes ("paradox of the faint sun") is probably related to a simultaneous change in the chemical composition of Earth's atmosphere that accompanies a fall of β.
2) The albedo has an influence on the earth's temperature. A rise in the albedo reduces the temperature. Actually, there were suggestions that, by increasing the albedo, the expected warming due to the anthropogenic Greenhouse Effect could be counterbalanced. A large mirror in the Earth's orbit, or additional particles in the lower stratosphere, would intensify the reflection of incoming solar radiation. Alterations in the size of snow-covered areas and cloud cover directly affect the albedo.
3) The surface characteristics of the Earth influence the thermal emission as well as the vertical transport of heat and moisture in the atmosphere. Thus, changes in surface characteristics, such as large areas of deforestation and urbanization, modify the discharge of energy from the surface. Such mechanisms were at the forefront of fears in the 19th century.
4) The ability of the earth's atmosphere to absorb long-wave radiation depends on the chemical composition of the atmosphere. Higher concentrations of absorbing substances in the atmosphere therefore lead to raised temperatures. Such substances are water vapor, carbon dioxide and also other gases, like chlorofluorocarbon or methane. The composition of the atmosphere has clearly changed through time. It is plausible that this change has balanced the above-mentioned rise in the output of the sun in the far geological past, so that the Earth has experienced only moderate temperature changes at that time. In the course of the Earth's history, absorbing gases in the atmosphere have become less abundant.

The Gaia Hypothesis (originally developed by James Lovelock [1919–] as the earth feedback hypothesis) was developed in this connection: according to this theory, the biosphere of the earth keeps actively natural environmental changes (caused, for example, by the sun) in check and thus makes possible a continuation of life on earth.

Humans have interfered with the climate since Neolithic times. Thus the transformation of Europe from a forested region into an arable one led at least to regional climate change. In modern terminology: it was an unintentional experiment in climate change effected by a change in the properties of the surface of the land (mechanism 3).

In 1794 Johann Gottfried Herder described this early influence of humans on the climate in picturesque and forceful terms:

"Since he stole the fire from heaven and learned to force the iron with his hand, since he became master of animals and their brethren and raised them as well as plants to be of service to him; he has in many ways contributed to changing the climate. Europe was formerly a moist forest and other now cultivated areas were equally large; (the forest) was thinned and along with the climate the inhabitants have also changed ... We can therefore consider the human race as a flock of bold, though small giants, who gradually climbed down from the mountains, subdued the earth and changed the climate with their weak hands. The future will teach us how far they will take it."

In North America too, such an "experiment" was conducted — the transformation of the prairie of the Midwest, large areas of forests in the East, and Florida's swamplands into arable land. In this case too, the changes cannot be described and analyzed by instrumental data. But experiments with climate models indicate that climatic changes were limited to the immediate region.

Today, two cases are discussed by the public; namely, the effect of the continuing deforestation of the tropical rain forests and the "anthropogenic Greenhouse Effect". We will not go any further into the question of rain forest deforestation here.

The Greenhouse Effect is a type 4 mechanism for possible causes of climate change; that is, it is based on a change in the chemical composition of the earth's atmosphere. As already explained, a certain portion of "radiatively active" gases in the earth's atmosphere is necessary in order to make possible a temperature hospitable to life. Today, however, the concentration of the radiatively active gases is dramatically raised, primarily by the burning of fossil fuels, so that a doubling of the carbon dioxide concentration may appear in a few decades. Methane emissions into the atmosphere, too, have risen in the last years and decades. Methane is emitted from rice fields and from domestic animals like cows, but also from the manufacture and transport of natural gas. The heating effect that results from these anthropogenic emissions is named the "anthropogenic Greenhouse Effect", which should not be confused with the natural Greenhouse Effect essential for life. Without CO_2 there is no photosynthesis, and therefore no plant world. To this extent, it is absurd to speak of carbon dioxide as a "climate killer" or "poison".

In the case of present emissions of carbon dioxide and other greenhouse gases continually rising, one expects a rise in the global air temperature of 2°C to 6°C by the 21st century compared to the "normal" 1960–90. At the same time, changes in precipitation distribution, as well as a rise in the water level of a few decimeters, up to 1 meter appear plausible. The anticipated changes will not occur by leaps and bounds, but will be gradual. Only many decades after the conclusion or after the stabilization of anthropogenic emissions do we expect an end to these changes. All statements about regional details are subject to great uncertainties.

Again and again the media express the fear that the polar caps[46] could melt, resulting in a rise of many meters in the water level. At the beginning of discussions in the late 1970s, individual scientists conjectured that the West Antarctic Ice Shelf was melting away and that a global water level rise of 6 m (or more) could be the result in the foreseeable future. Today no seriously regarded climate researcher says that any more, although certainly journalists, environmental organizations and advocate scientists sometimes do. On the contrary, climate

---

[46] "Polar caps" actually refer to the Antarctica and Greenland, not the icy cover of the North Pole, which is made up of swimming sea ice, the melting of which would be inconsequential for sea level.

researchers maintain positively that it is possible that the ice caps of Greenland or the Antarctic could grow, due to enhanced precipitation. And rainfall transforms itself into ice, independently of whether it falls at a temperature of −30°C or of −25°C.

Simulations with climate models support these expectations. In these models, the chemical composition of the earth's atmosphere is steadily changed by CO_2 emissions, which typically increase by 1% per year. This assumption of a 1% increase is not trivial, and is recommended by economists as plausible. However, considering the unsteady and unpredictable development of the global commercial system over long periods of time, the assumption is certainly questionable. Occasionally, simulations are calculated with diminished emission rates, such as a future stabilization or reduction in an emissions level from an earlier year. In each case, all calculations indicate that a stabilization of emissions at today's level will lead to a stabilization of the temperature, at a rather high level, only after several decades. Keeping today's temperature will only be possible — according to these calculations — if today's emissions are strongly reduced. Figure 22 shows several hypothetical emissions scenarios and the "answers" of the climate system in terms of atmospheric CO_2 concentration, the air temperature and the mean sea level as determined by the Intergovernmental Panel on Climate Change (IPCC) in its Third Assessment Report[47] from the work of various research groups and models.

There are several comments needed when discussing Fig. 22.

1) First, the emissions are scenarios that are considered possible, plausible, and internally consistent, but not necessarily probable. The idea is to "span" the range of possibilities, to describe different futures, so as to allow societies to ponder perspectives and options. As we will describe below, these scenarios depend on a variety of assumptions, which cannot be predicted (in the sense of specifying a most probable future with given timing).

---

[47] The IPCC's position on these issues has hardly changed since 2001, even if by now a Fourth Assessment has been published.

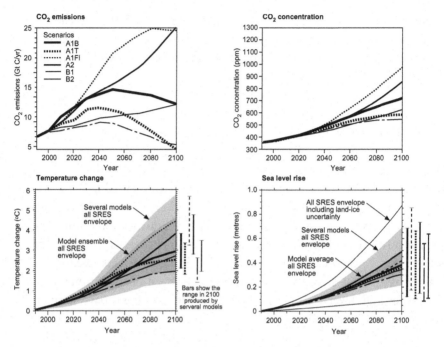

Fig. 22. IPCC scenario of possible, plausible and physically consistent developments of carbon dioxide emissions, resulting concentrations of carbon dioxide in the atmosphere, global mean air temperature at the surface and global mean sea level. The stippled areas indicate uncertainty ranges.

Source: The Third Assessment Report of the IPCC (2001).

2) The second comment is related to this uncertainty — the description of possible future temperature and sea level conditions are also only scenarios — possible, plausible, and internally consistent[48] but not necessarily probable futures. They are not predictions in any strict sense. They depend on the assumed emissions, and thus they are technically "conditional predictions", which is a very different thing from a "prediction" in common language. In many quarters, unfortunately,

---

[48] "Internal consistency" is a non-trivial requirement. The movie *The Day After Tomorrow* was seen by some as a scenario. It was not, as it described a future, which is physically impossible. The movie provided only a story, which was either badly researched or optimized for best entertainment.

the unqualified term "prediction" is used. This is simply sloppy language and should be avoided.
3) The developments of emissions deemed possible are very different. Interestingly, the resulting atmospheric concentrations deviate from each other much less; eventually, the temperature and sea level variations are also not that different. Until about 2040, the changes of temperature and sea level are almost identical, irrespective of the emission scenarios.
4) To avoid a marked increase in temperature and sea level requires massive reductions of emissions.

In the following we will look at the emissions scenarios in some more detail, so that the reader gets an impression which issues have been touched upon when constructing such scenarios. Key emission scenarios have been published as the "IPCC Special Report on Emissions Scenarios",[49] prepared by economists and other social scientists for the Third Assessment Report of the IPCC. They utilize scenarios of greenhouse gas and aerosol emissions, or of changing land use.

(A1) A world of rapid economic growth and rapid introduction of new and more efficient technology.
(A2) A very heterogeneous world with an emphasis on family values and local traditions.
(B1) A world of "dematerialization" and introduction of clean technologies.
(B2) A world with an emphasis on local solutions to economic and environmental sustainability.

The scenarios do not anticipate any specific mitigation policies for avoiding climate change. The authors emphasize that "no explicit judgments have been made by the SRES team as to their desirability or probability".

---

[49] SRES, http//www.grida.no/climate/ipcc/emission. See also R. S. J. Tol; Economic scenarios for Global Change, in H. von Storch, R. S. J. Tol and G. Flöser (eds.), *Environmental Crisis. Science and Policy* (Springer Verlag, 2008).

The Scenarios A2 and B2 have been widely used in recent years. Therefore, we explain the socio-economic background of these scenarios in more detail:[50]

SRES describes the A2 scenario as follows:

> "... characterized by lower trade flows, relatively slow capital stock turnover, and slower technological change. The world "consolidates" into a series of economic regions. Self-reliance in terms of resources and less emphasis on economic, social, and cultural interactions between regions are characteristic for this future. Economic growth is uneven and the income gap between now-industrialized and developing parts of the world does not narrow. People, ideas, and capital are less mobile so that technology diffuses more slowly. International disparities in productivity, and hence income per capita, are largely maintained or increased in absolute terms. With the emphasis on family and community life, fertility rates decline relatively slowly, which makes the population the largest among the storylines (15 billion by 2100). Technological change is more heterogeneous. Regions with abundant energy and mineral resources evolve more resource-intensive economies, while those poor in resources place a very high priority on minimizing import dependence through technological innovation to improve resource efficiency and make use of substitute inputs. Energy use per unit of GDP declines with a pace of 0.5 to 0.7% per year. Social and political structures diversify; some regions move toward stronger welfare systems and reduced income inequality, while others move toward "leaner" government and more heterogeneous income distributions. With substantial food requirements, agricultural productivity is one of the main focus areas for innovation and research, development efforts, and environmental concerns. Global environmental concerns are relatively weak."

In B2, there is

> "... increased concern for environmental and social sustainability. Increasingly, government policies and business strategies at the national and local levels are influenced by environmentally aware citizens, with a trend

---

[50] For a summary for the other two scenarios, refer to Müller and von Storch, *op cit.*

toward local self-reliance and stronger communities. Human welfare, equality, and environmental protection all have high priority, and they are addressed through community-based social solutions in addition to technical solutions. Education and welfare programs are pursued widely, which reduces mortality and fertility. The population reaches about 10 billion people by 2100. Income per capita grows at an intermediate rate. The high educational levels promote both development and environmental protection. Environmental protection is one of the few truly international common priorities. However, strategies to address global environmental challenges are not of a central priority and are thus less successful compared to local and regional environmental response strategies. The governments have difficulty designing and implementing agreements that combine global environmental protection. Land-use management becomes better integrated at the local level. Urban and transport infrastructure is a particular focus of community innovation, and contributes to a low level of car dependence and less urban sprawl. An emphasis on food self-reliance contributes to a shift in dietary patterns toward local products, with relatively low meat consumption in countries with high population densities. Energy systems differ from region to region. The need to use energy and other resources more efficiently spurs the development of less carbon-intensive technology in some regions. Although globally the energy system remains predominantly hydrocarbon-based, a gradual transition occurs away from the current share of fossil resources in world energy supply."

Expected emissions of greenhouse gases and aerosols into the atmosphere are derived from these assumptions and descriptions. Figure 22 shows the expected SRES scenarios for carbon dioxide (a representative of greenhouse gases; in gigatons per year).

The SRES scenarios are not unanimously accepted by the economic community. Some researchers find the scenarios internally inconsistent.[51]

---

[51] See R. S. J. Tol; Exchange rates and climate change: An application of FUND, *Climatic Change*, 2007, and House of Lords, Select Committee on Economic Affairs, *The Economics of Climate Change*, Vol. I: Report, 2nd Report of Session 2005–06. The Stationery Office Limited, HL Paper 12-I, *Authority of the House of Lords, London, UK*, 2005. (http://www.publications.parliament.uk/pa/ld/ldeconaf.htm#evid)

A key critique is that the expectation of economic growth in different parts of the world is based on market exchange ranges (MER) and not on purchasing power parity (PPP).[52] Another aspect is the implicit assumption in the SRES scenarios that the difference in income between developing and developed countries will significantly shrink until the end of this century. These assumptions, the argument is, lead to an exaggeration of expected future emissions.

So much on the emissions scenarios.

Naturally, the question is whether the first signs of the expected global climate changes can now be observed. This question is frequently abbreviated in the public's mind to the interpretation of events of short duration, such as one or a few consecutive cool summers, or a series of heavy storms. Actually such events are "normal" in the sense of natural climate variations: the chances of a hundred-year rainfall, storm or cold spell at your actual location are small, but the probability that somewhere a hundred-year occurrence will take place is not that small. In other words: the probability that a person hits all the numbers in the weekend lotto is extremely small; but the probability that someone really will have all of them is almost 100%.

A convincing "detection" of anthropogenic climate change can only succeed when long series of observation data are available. Only if they include many decades is discrimination between "normal" and "abnormal" development possible. In order to recognize anthropogenic climate change, the suspected variation must be larger than naturally caused changes, or different in its spatial structure. If one looks at hurricane damage statistics for the US coasts in Fig. 7 over the last 30 years, one finds a constant rise. But if one compares these numbers with (homogenized) damage data from the beginning of the century, one finds similarly high values to those in recent years (Fig. 8). Therefore, data covering less than 30 years is insufficient to judge the normality or the deviation from naturally occurring patterns. Many interpretations of systematic changes are based on impermissible generalizations from analyses of data sets that are too short. This is why satellite data are mostly useless for detection purposes.

---

[52] A general critique of the social dynamics behind this process is provided by A. Kellow; *Science and Public Policy* (Edgar Elgar Publishing, 2007).

Also, the data must be homogeneous throughout, that is, changes in the data must be based on modifications in the environment, and not on modifications of the observation procedure. Further, over the observation period they must be representative for the same area. The storm data in Fig. 4 are not homogeneous, because they do not represent a change in the frequency of storms, but rather the difference in wind speed in two different places in Hamburg. The same is true for satellite data, which may point to false developments because the geometry of the satellites' orbits alter slowly, and discontinuities in the orbit appear when old satellites are replaced by new ones. The temperature series from Sherbrooke in Fig. 6 is in this sense likewise unsuitable, not only because of the discontinuities in the data, but also because of the urbanization effect. For reasons of this sort, much observation data cannot be used for analysis of actual climate trends. As a rule of thumb, only the series of air temperatures over land and sea (measured by ships and at rural meteorological land stations), as well as air pressure data series, fulfill the necessary requirements of sufficient length and homogeneity.

An analysis of the global temperature data sets actually shows that for approximately a hundred years, with some interruptions, the air temperature has risen. The warming trend of the last 30 years has a spatial distribution pattern anticipated by climate models. It is larger than all trends that were found in observations until now, or in climate simulations that were conducted without elevated greenhouse gas concentrations. Including all observed and simulated trends in a probability distribution, show that the probability of finding a trend like the last one observed without a causal connection to the unquestioned changes in the chemical composition of the earth's atmosphere amounts to much less than 5%.[53] In the terminology of statistics, this means: "significant at the 95% level". In the determination of this significance level, the important proviso is that the estimate of natural variability is correct. Besides, some media sources transform this idea of "95% significance"

---

[53] This statement is somewhat imprecise: "...without external factors at work". In other words; natural processes probably do not generate the changing temperature pattern; it is plausible that the changing chemical composition of the earth's atmosphere is at least partially the cause for the rise in temperature.

into absurd statements as: "95% of the warming is of anthropogenic origin".

At the end of the 1980s, the "Intergovernmental Panel on Climate Change" (IPCC) was formed to consider the scope of climate change. The experts comprising the IPCC issued four detailed reports (1990, 1996 and 2001 and 2007) summarizing the state of research concerning anthropogenic climate change. It is considered to be irrefutable that the concentration of radiatively active gases has dramatically increased since the beginning of industrialization. The expected implications for the future of this elevated concentration come from calculations by means of climate models, because the observation data is too short, too inhomogeneous and (because of natural climate variability) affected by natural variability, sometimes called climatic noise. In the early 1990s, the recent increases of the surface temperature averaged over wide parts of the globe and over several years appear dramatic *per se*, but they were only slightly stronger than the rises in temperature in the 1920s and '30s. Therefore, in 1990 the IPCC still cautiously stated: "... *the size of this warming is broadly consistent with predictions of climate models, but it is also of the same magnitude as natural climate variability. ... The unequivocal detection of the enhanced greenhouse effect from the observations is not likely for a decade or more*".[54] In the report of 1995 it was stated more clearly: "... *the balance of evidence suggests a discernible human influence on global climate*".[55] In 2001 and 2007, the formulations were considerably less cautious: "*there is new and stronger evidence that most of the warming observed over the last 50 years is attributable to human activities*" (2001) and "*Advances since the TAR show that discernible human influences extend beyond average temperature to other aspects of climate, including temperature extremes and wind patterns*" (2007).

By now, the "detection" of anthropogenic causes of the recent warming is considered "proven" by a wide margin of majority in the scientific

---

[54] J. L. Houghton, G. J. Jenkins and J. J. Ephraums (eds.). *Climate Change. The IPCC scientific assessment* (Cambridge University Press, 1990), p. xii.

[55] J. T. Houghton, L. G. Meira Filho, B. A. Callander, N. Harris, A. Kattenberg and K. Maskell (eds.). *Climate Change 1995. The Science of Climate Change* (Cambridge University Press, 1996).

community. In fact, a simple exercise demonstrates that we are presently experiencing a development that would be extremely rare if no human influence would be present: at the time of the Fourth Assessment report of the IPCC in 2007, global mean temperatures were available since 1880, i.e., the last 126 years. Among these were the 13 warmest years in the last 17 years, i.e., since 1990. How probable is such an outcome if we had stationary conditions (which would be the case without external influences)? If the global mean temperatures represented independent and statistically identical events, then the probability of such an event would be 1.25×10^{-14}. However, there is memory across the years; if we take this properly into account, we end up with a probability of 0.1% or less.[56]

4.5. Climate Change as a Social Construct

The common understanding of climate and climate change has barely been investigated until now. Analysis of the everyday comprehension of climate is certainly no easy task. In particular, the question of the complex social construct of climate and weather and its origins is masked by the routine practice with which we commonly use this concept. This custom suggests that we have something almost like "natural" intuitive insight into and taken-for-granted attitude toward these phenomena.

However, the everyday presence of these themes also conceals a certain ambiguity, fragility and perhaps even misunderstanding of climate in everyday life. Lay people have deeply rooted conceptions of climate. Knowledge of common interpretations of climatic conditions is not only of interest because these interpretations offer insights into the societal perceptions of natural processes; they are also significant because in certain cases they can provide important indications of how the public reacts to scientific knowledge concerning climate and climate change. Finally, this common understanding of climate influences each conception of, and each reaction to, climate politics.

---

[56] E. Zorita, T. F. Stocker and H. von Storch, How unusual is the recent series of warm years? *Geophys. Res. Lett.*, Vol. 35 (2008).

Eduard Brückner gives us a crucial hint, perhaps, when he says:

"The assuredness of the invariability of climate is deeply rooted in the people and expresses itself in the certain conviction that the unusual weather of one season or year will be compensated for in the following one."[57]

Apparently people in Western societies have almost always believed, at least since the Enlightenment, that weather was getting worse. In 1996, the Swiss geographer and climate scientist Martine Rebetez examined the case of "white Christmases".[58] It is generally held as a truism that today there are seldom "white Christmases", while in earlier times snow on Christmas was the rule. The evaluation of the available observation data for a series of weather stations in Switzerland produces a probability for snow on Christmas Day in Zürich amounting to only 25% (or, in other words: only one out of four Christmases is, on average, white). For higher lying places such as the town of Einsiedeln, located at 850 m altitude, the probability is over 80%; while in the city of Geneva, at 400 m altitude, it is somewhat less than 20%.

Rebetez' significant conclusion is that the probability for snow over the course of time has not decreased at the time of publication of the article in 1996. In fact, there has been even somewhat more snow in the last century and not less. That is, the public is correct in its opinion that the probability for Christmas snow is small; but it is not correct that this is somehow unusual. It is much more the rule, and not the exception, that at the end of December there is no snow in low-lying places in Switzerland.

As an aside, Dr. Rebetez also examined the question of where the powerful mental model of White Christmas may originate from. To do so, she examined Christmas cards, and noticed that only at the end of the 19th century did these cards show St. Claus on a snow-covered roof, a habit imported from New England, while previously the roofs had been bare. An example is provided by Fig. 23.

---

[57] E. Brückner, Klimaschwankungen seit 1700. Nebst Bemerkungen über die Klimaschwankungen der Diluvialzeit (Wien; Hoelzel, 1890).

[58] M. Rebetez, Public expectation as an element of human, perception of climate change, *Climatic Change* 32 (1996) 495–509.

The American anthropologist Willett Kempton and his colleagues have conducted a study about the everyday understanding of climate, climate changes and threats from climate among American lay people.[59] First of all, a group of 20 lay people was interviewed with open-ended questions; then a questionnaire was developed that was answered by a larger group of people, in order to test to what extent the opinions originally

Fig. 23. Christmas card motifs — an older one showing St. Claus on a bare roof (1845), and a newer one (1863) on a snow-covered roof.[60]

---

[59] W. Kempton, J. S. Boster and J. A. Hartley, *Environmental values in American Culture* (MIT Press, Cambridge MA and London, 1995).
[60] Rebetez *op cit.*, pp. 500–501.

expressed by individuals in the interviews were typical of individual beliefs.

One question was: "Which factors would you say affect the weather?" Some answers were consistent with the meteorological school of thought; namely, "jet stream" (one of the factors rightly emphasized repeatedly by television newscasts) or "sunspots, volcanic eruptions, geological processes", while other answers were surprising:

1) "Pollution affects the weather."
2) "Burning, like these rain forests and these Western forest fires. Spraying from insecticides and stuff like. And herbicides, like on the farms, to prevent weeds from growing …. But the most important thing is burning and auto pollution."
3) "… that bomb I think it had an awful bearing on our weather. The A-bomb. They had those tests … just seemed like here things have changed ever since; it's become more torrent, the weather here in the past few years … the weather is very changeable."
4) "…my own private theory … that every time they shoot something up in the space it disturbs things up here! … It just seems every time something happens we get this strange type of weather… tornadoes were rare in this section of the country … it used to be rather calm here …"

Finally, the researchers asked their subjects whether they agreed with the statement: "there may be a link between the changes in the weather and all the rockets they have fired into outer space". 43% of the respondents agreed.

Even when almost all those questioned could conceive, in some fashion, of such concepts as "Global Warming" or "Greenhouse Effect", most had mental models that deviated significantly from scientific understanding of the same processes.

The most frequent lay mental model is "environmental pollution" after the example of "acid rain". The damaging substances were viewed as artificial, poisonous for living organisms: "Well, I like warm weather personally, but I think it's wrong what humans are doing to the atmosphere … with all the aerosols and the ozone and so forth … it's wrong because at the same time, we are ingesting and breathing in all these different chemicals that are being put into the atmosphere".

The sources of emissions of these damaging materials were perceived to be mainly cars and industry. According to this assumption, the problem can be solved by the introduction of suitable filters. But this model is wrong, because the emitted matter, namely carbon dioxide — a common, naturally occurring gas — is not harmful to the health, and is exhaled by each of us all the time. The main sources of carbon dioxide are combustion processes in power stations, traffic and heating. There is at present no commercially available filter for limiting the emission of CO_2 into the atmosphere; the only efficient option at this time is decreasing the use of fossil fuels through conservation or more efficient use. However, techniques are now being developed that should filter out existing CO_2 in the waste air of power stations.

The problem of "Global Warming" is often confused with the problem of ozone destruction in the stratosphere. Thus, a respondent says: "... the amount of protective atmosphere around the world ... these layers are getting thinner and thinner, and as more heat gets through that tunnel ..." In newspapers the talk is of: "Ozone-destroying carbon dioxide emissions". Actually both processes, "anthropogenic Greenhouse Effect" and "stratospheric ozone reduction", are linked together, because the ozone-destroying CFCs also contribute to the Greenhouse Effect. But "Global Warming" is above all about carbon dioxide, which is not affecting the stratospheric ozone and does not influence the ozone hole.

A third model results from the facts that the quantity of oxygen and carbon dioxide are directly connected and that a rise in the concentration of carbon dioxide must lead necessarily to a decrease in concentration of oxygen: "Pretty soon we won't have any more oxygen to breathe". According to this view, the rise in concentration of carbon dioxide is mainly brought about by the deforestation of woodlands, because the trees transform the carbon dioxide produced by man and animals back into oxygen. The assertion: "If they cut all the forests down, we will soon run out of oxygen" was agreed upon in 77% of all questionnaires. In fact, a simple rough estimate shows that a burning of all forests (with a size of about 5,000 giga-tons of carbon) would reduce the atmospheric concentration of oxygen of today, from 20% to 19.8%.

To sum up, one can say that the American interviews revealed scientifically inadequate conceptions by lay people with different professional backgrounds, from environmental activists to loggers unemployed due to environmental politics, but also by congressional advisors. Even though the survey was carried out more than 15 years ago, we are convinced that the major conclusions still hold.

How did these common sense ideas originate? Which factors and forces participate in the process of the social construction of climate knowledge and the public understanding of climate? To date this area has still not been systematically explored; but it would certainly be interesting and would yield useful results for the political sphere. Presumably there are a series of important factors:

1) Traditional conceptions of climate and climate change, as we have discussed in the preceding section. It was a topical theme at the turn of the millennium when fundamentalist preachers, at least in North America, announced the biblical end of time to their television audience congregations. Climate catastrophes and extreme weather occurrences fitted very well here.
2) Interpretation of new developments with the help of ideas that agree with other environmental problems. Examples are the already mentioned acid rain and the stratospheric ozone depletion.
3) Sensationalized reports in the media and representations in popular science books show a clear trend toward using exaggerations for the purpose of increasing circulation. Also, these reports grossly generalize factual details, while failing to explain the crucial ones. In the following we present a few examples from recent years:

The cover of an English book dramatizing the climate problem states:

"We shall be engulfed, quite literally, by the consequences of our greed and stupidity. Nearly two-thirds of our world could disappear under polar cap water, melting as a result of ozone depletion and deforestation."

Here they explicitly haul out the big guns of melting polar caps. As already mentioned, there is no plausible scientific argument for such a meltdown. This conjecture has nothing to do with ozone depletion or deforestation.

— In June 1994 the highly regarded Danish newspaper *Politiken* wrote:

> "The environmental organization Greenpeace has published a report about 500 extreme weather events — hurricanes, record temperatures, droughts and similar occurrences — from the last three years. These extremes have increased lately and were understood by Greenpeace as the first signs of the Greenhouse Effect. The report, "The Time Bomb," that was delivered to the Minister of the Environment should be brought up every six months."

As we have seen, with data covering only three years, useful assessments about natural and unnatural climate variations are not possible.

— A member of the German Federal Parliament, publicly considered a climate expert, explained in the 1990s in a German newspaper:

> "The changes in the climate system — especially the increase in extreme variations and unusual meteorological phenomena — are without doubt due to human beings."

Doubtless the insured damages caused by storms and other weather phenomena clearly rose in the past decades, but it is by now clear that altered lifestyles and resource utilization have been the major factor behind rise in damage insurance numbers.[61] A causal connection between occurrences of extremes is so far scientifically unproven (apart from the trivial effect that when it gets warmer in general, the hot days will be more frequent and the cold days less frequent).

What is interesting about all of these representations and reconstructions of scientific knowledge is that the reworking and filtration of the scientific knowledge creates a new reality. Later, science finds itself in the

---

[61] See the consensus report on the Workshop on *Climate Change and Disaster Losses: Understanding and Attributing Trends and Projections*, available on http://sciencepolicy.colorado.edu/sparc/research/projects/extreme_events/munich_workshop/executive_summary.pdf.

strange situation that its success or failure is assessed and adjudicated by comparing the development not against the original scientific findings, but rather against the published metamorphosis.

For example, an article, "The Moods of the Sun", appeared in the German weekly DIE ZEIT in July 1997, criticizing climate science scenarios of climate change because the observational evidence did not support the published scenarios, which had been exaggerated by the media. The climate researcher Klaus Hasselmann, in his reply "The Moods of the Media", called attention to these cycles, which always give the media the benefit of new news; at the cost, however, of failing to convey objectively the facts to the public. In the beginning, the scientific findings are exaggerated, with a positive impact on circulation; later the discrepancy between the scientific and published findings is used to discredit science — this again furthers circulation.

Naturally, commercial and socio-political interests are also in play. Because energy production from fossil fuels stands at the center of the climate problem, competition and conflicts occur. Environmental movements employ the anthropogenic "climate catastrophe" as a striking example of the exploitation and misuse of nature by irresponsible industrialized societies, with terrible consequences for people and the ecosystem. Insurers see their market chances improved when the perception of increased risk prevails among their clients and in the public.

Climate scientists, too, sometimes play a dubious role in the context of the public career of environmental themes. Next to the duty to provide information, there is also another, possibly unconscious motive to go public. Such motives can be the prospect of more research funding, a general desire to better the world, or simply the pleasure of seeing oneself in the media spotlight.

Scientists also know that dramatized representations get more attention from the public and political decision-makers; the readiness to "pay attention" improves. That means that the media strive to use certain rhetorical strategies not only in order to be heard, but also to be convincing, in a public discourse with many competing topics. Interviews with climate researchers in the media usually begin with a presentation of scientific findings, but then the journalist asks the scientist about the consequences for the general public, economy or politics. At this moment the scientist relinquishes his expertise and begins to speculate as an educated layperson about complex societal relations.

Typical for this scheme is the following extract from an interview in a German newspaper:

> "And what will happen with the atmosphere?" — "We have to expect more and more frequent strong low pressure systems and storms. Perhaps then agriculture can no longer be practiced as in the past, because with the rising sea level, salt water may intrude in our ground water. Also, the Sahara could, for example, spread over the Mediterranean. And if certain stretches of land are no longer habitable, people will be drawn toward places where still acceptable conditions prevail. There would be migrations and climate wars."

Interviews with climatologists who decline to respond to questions about societal consequences of climate change because they lie beyond their discipline-specific competence, will sometimes not be broadcast at all.

In 2003 climate researchers — chiefly in the U.S.A. and Europe, but also in other parts of the world — were asked for their opinions using a questionnaire.[62] Among the questions were the following:

1) How much do you feel that scientists have played a role in transforming the climate issue from being a scientific issue to a social and public issue?
2) Some scientists present the extremes of the climate debate in a popular format, with the claim that it is their task to alert the public. How much do you agree with this practice?

The respondents answered on a scale of 1 to 7, where a 1 indicates strong agreement with the statement and a 7 indicates a clear rejection. A 4 represents an undecided position. The results are shown in Fig. 24 as frequency distribution.

---

[62] For details, refer to D. Bray and H. von Storch, Climate Scientists' Perceptions of Climate Change Science, GKSS-Report 11, 2007. http://coast.gkss.de/staff/storch/pdf/GKSS_2007_11.pdf.

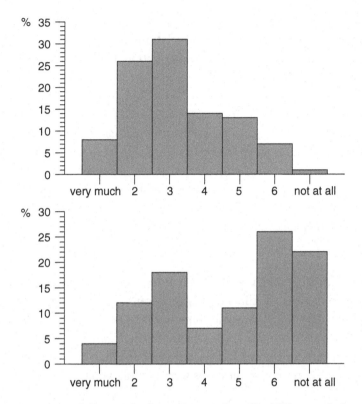

Fig. 24. Responses of climate scientists to the questions: (Top) "How much do you feel that scientists have played a role in transforming the climate issue from being a scientific issue to a social public issue?" (Bottom) "Some scientists present the extremes of the climate debate in popular format with the claim that it is their task to alert the public. How much do you agree with this practice?"

The respondents agree with the observation that scientists themselves have played a role in the transfer of the climate problem from the scientific into the political arena. The answers to the second question are divided. On one side the largest group of scientists (categories 5–7; sum about 71%) decries such an attitude. Another, smaller group of about 20% (categories 1–3) support extreme representations, with the aim of alarming and warning the public. In this group only very few, namely 4%, are very much in favor of such practice, while the other 16% have some reservations.

To summarize: Through global warming, the destruction of the ozone layer, deforestation of forests, modern methods of transport and similar processes, the conception of a climate change caused by mankind in recent times has emerged, which causes controversial public discussions. While the history of civilization until now has been understood as one of the emancipation of society from nature (including climate), perceptions have arrived at a dramatic turning point, in which nature once again increasingly rules people. As a punishment for humanity's playing with the ecological balance, nature becomes sick and causes sickness: "Nature strikes back". The question is naturally whether nature itself changes or, as a result of our research efforts, our manner of perceiving nature changes.

This change is often described with the help of terminology borrowed from medicine, for example, the term "syndrome" for the diagnosis of characteristically pathological environmental situations. The treatment of this syndrome demands diagnosis and prescriptions for treatment, chiefly by scientific system analysts.

It is scientific knowledge that sets in motion and structures political processes. Science formulates the problem of climate change for politics and society: the discovery of global climate change, of the Greenhouse Effect and rise in temperature is no everyday problem. It is the scientific formulation of problems that determines to a large extent the kind and extent of political consequences. Scientists know this (compare the first question in Fig. 24). They play a special role in forming and changing the everyday understanding of climate. In doing so, they also pursue political, ideological and other subjective interests alongside their scientific interests. To that end, scientists leaning toward drastic formulations have sufficient opportunity in certain media. They find support by vested interests, such as journals interested in circulation numbers or companies selling protection from risk or measures to reduce one's carbon footprint.

It should be stressed, however, that scientists can hardly be considered to be spreading a message of their choice. Instead, the message of manmade climate change, which influence the world for the worse, is consistent with the social constitution of climate and climate change. It is the coincidence of scientific insights and of cultural knowledge claims that makes the issue of climate a particularly potent agent for the public and political arena, at least in the west.

In the following section, we will document in some detail how this mixture of knowledge claims by authorities, namely originally religious leaders and later leading members of the scientific community, with their preconceived knowledge of humans either destroying the Creation or completing the Creation, has resulted in a long series of cases of perceived anthropogenic climate change.

4.6. The History of Anthropogenic Climate Catastrophes

The notion of anthropogenic climate changes is not an idea formulated and found in contemporary discussions only. The following list of perceived or expected anthropogenic climate catastrophes is certainly incomplete; but it includes not only examples of the everyday and frequently observed fear of technical innovations, but also references to impending climate catastrophes and their causes given to us by science. Likely our list is far from complete, as it is based more on coincidence than on systematic studies.[63]

A frequent case of alleged climate catastrophe concerns the interpretations of extreme weather regimes or climatic events spread by religious leaders. Events of this sort were and are interpreted as God's punishment for human sins. An example is the case, discussed below, of the English famine in the years 1314–1317. Another case of climate changes understood as anthropogenic concerns the previously discussed acts of medieval witches. Besides direct weather magic, such as a harvest-annihilating hailstorm, there was still the indirect mechanism: the possibility that God Himself showed his anger at insufficient action against local witches and punished the negligent community by extreme weather. Thus the idea that immoral behavior could shift the equilibrium of the climate has been with the human race presumably since its earliest beginnings.

The oldest scientifically documented case known to us concerns the climatic effects of the cultivation of the North American colonies in the 17th and 18th centuries. Doctor Hugh Williamson reported in 1770 on climatic

---

[63] H. von Storch and N. Stehr, Climate change in perspective. Our concerns about global warming have an age-old resonance, *Nature*, Vol. 406 (2000), p. 615.

changes in the then English colonies.[64] He stated that the climate in New England had improved, and this was due to the altered use of land; particularly the terrible northwest storms in winter had lessened. This is one of the few cases where human activity is credited with bettering climate.

At the end of the 18th century and the beginning of the 19th century, assertions were disseminated in Germany and Switzerland, according to which alterations in precipitation were caused by lightning conductors. Authorities found it necessary to contradict these assertions and to warn urgently of the possibility of forcible action.

Thus, the *Neue Zürcher Zeitung* of 9 July 1816 states:

"On 30 June this proclamation was read out from the pulpit: 'The local authorities ... are ordered ... to make the people understand, how regrettable and awkward is was to the government, to become aware not only that the erroneous illusion and false prejudice existing today have spread in several counties that the protection of buildings with the erection of lightning conductors against lightning strikes causes weather detrimental to agriculture, but also that some highly irrational or wicked men took the occasion to endanger the public security and sacredness of property by menacing or attempting forceful destruction of groundlessly suspected lightning conductors ... But should the evil-minded and disturbers of the peace with their foolish and absurd allegations of harmfulness against the lightning conductors wrongly seize the property of their fellow citizens, disturb the peace of their households and endanger the public security, so ... the most exact orders are given to the local authorities by the high government, to avenge and punish each outrage and each misdeed, as a deterrent example for others, by use of the strictest laws."

In the introduction to his book *Climate Changes since 1700*, Eduard Brückner summarizes the discussion of natural and anthropogenic climate changes in the 19th century. Accordingly, the altered use of land, particularly

---

[64] H. Williamson, An attempt to account for the change of climate, which has been observed in the Middle Colonies in North America, *Trans. Amer. Phil. Soc.* 1 (1770) 272.

deforestation and afforestation, was to be understood as a powerful climate-altering factor:

> "In the early 1870s Wex published his well-known work about the reduced amounts of water in springs, rivers and streams. ... Wex concluded that in cultivated areas, decreasing water levels result in a continuous decline in precipitation ... general conclusion: a continuing decrease of the water levels of springs, rivers and streams takes place in cultivated areas, caused mainly by the increased practice of deforestation and its resulting decrease in rainfall. This proof had raise some serious concerns. In 1873, in Vienna, the Congress for Agriculture and Forestry discussed the problem in detail; and when the Prussian House of Representatives ordered a special commission to examine a proposed law pertaining to the preservation and implementation of forests for safeguarding, it pointed out that the steady decrease in the water levels of Prussian rivers was one of the most serious consequences of deforestation, only to be rectified by reforestation programs. It is worth mentioning that at the same time or only a few years earlier the same concerns were raised in Russia as well, and governmental circles reconsidered the issue of deforestation."

The atomic bomb tests in the 1950s and 1960s were again and again linked with lasting climatic consequences. The *New York Times* of 8 July 1962 announced under the headline:

> "The weather — many human activities change it, but mostly not for the better": "If there is one thing that farmers in my area agree about, it is that the weather isn't the way it used to be any more. It's gotten worse. They'll say to you that summer is stormier, winter longer and spring later. And extensive agreement also persists with regard to the causes for these apparent changes: The bomb has done it."

This conception was also expressed repeatedly in the above-mentioned interviews with American respondents conducted by Kempton and his colleagues.

Even about 100 years ago, proposals were made to divert large Siberian rivers in Central Asia to intensify agricultural land use and prevent a draining of the Aral Sea. After the 25th Party Congress of the Communist Party of the Soviet Union in March 1976, planning for such an undertaking began in earnest. Besides ecological protests, such as with regard to fish stocks, concerns about climatic changes were voiced. There were fears of alterations in the distribution of sea ice in the Arctic Ocean as a consequence of the decreased freshwater discharge of the rivers. Such changed sea ice conditions would then, for their part, influence the climate in the whole Northern Hemisphere. Another more optimistic vision was that less sea ice would lead to milder conditions in Siberia. The plans were not realized, so that one cannot say to what extent the warnings and visions then were warranted. But model calculations indicate that the climatic effects of the planned manipulation of the Siberian rivers would have been rather moderate.

The progressive deforestation and burning off of tropical rain forests is commonly understood today to be a definite endangerment of the global climate, although the climatic effects have only a regional character, just like the transformations of the North American prairies in the past.

The vapor trails of high-flying airplanes are also frequently suspected of having a damaging climate effect. Current and future air traffic also contributes to the Greenhouse Effect today, even if the present contribution is still relatively small. But, as an expert points out, this could change, because the "Air traffic per unit fuel mass affects the climate more seriously than that of other traffic systems", and "the fuel consumption of air traffic grows more sharply than most other anthropogenic CO_2 sources".

At the height of the Cold War, deliberations took place about which climatic effects would be associated with a nuclear war. Generally widespread large surface fires were anticipated as a consequence of large disturbances on the ground. The fires, it was thought, would inject smoke particles in the atmosphere. A portion of these smoke particles would reach the stratosphere and remain there, much like volcanic aerosols, for many months, thereby effectively impeding the solar radiation from reaching the Earth's surface (like the "year without summer" after the eruption of the volcano Tambora). Naturally this would be linked to the worst consequences for the earthly biosphere, and life as presently known would become impossible.

A regional variant of "nuclear winter" began in connection with the invasion of Kuwait by Iraq in 1991. The Iraqis threatened to set fire to Kuwait's oil wells in case of an American attack. Scientists warned that the smoke from such fires would rise into the stratosphere and could blank out sunlight even on a large scale. The summer monsoon, an indispensable climatic feature for farmers in India, could fail to appear, and catastrophic famines might follow. When the Iraqis made good on their threat and set fire to the oil wells in Kuwait, the smoke rose to an altitude of a few thousand meters, certainly without reaching the stratosphere. The worst environmental damages occurred within a circumference of a few hundred kilometers, but the feared long-range effect failed to appear.

A relatively new concept of possible anthropogenic climate changes has the Gulf Stream as its theme, more precisely the "breakdown of the Gulf Stream". The warm Gulf Stream flows along the U.S. American East Coast to the north and then from Cape Hatteras northeasterly across the Atlantic, so that in Northern Europe a comparably milder climate rules than in other areas at equally northern latitudes, such as, for example, Alaska. With the disappearance of the Gulf Stream, the influx of warmth ceases and in northern Europe cold periods, perhaps even ice ages, would begin. This development is considered a conceivable by-product of global warming, so that together with the global warming regional cooling might appear. Experts maintain that such a development is improbable as long as the CO_2 concentrations do not grow to the extent of multiplying the current value.

A somewhat strange conceptual experiment was published in the summer of 1997 in *Transactions of the American Geophysical Union*. It was based on the observation that in the Mediterranean Sea, enormous amounts of water evaporate, so that the Mediterranean's water is very salty. This salty and thereby heavier water leaves the Mediterranean at the bottom of the Straits of Gibraltar, while on the surface less saline Atlantic water fills up the Mediterranean again. The article asserts that this cycle is accelerating because less fresh water is discharged into the Mediterranean — above all because of the Aswan Dam in Egypt — and because in addition the evaporation is intensifying due to the anthropogenic Greenhouse Effect. The saline heavy water entering the Atlantic would retard the Gulf Stream and finally heat the Labrador Sea, so that increased moisture will be carried into Canada. This moisture precipitates in Canada chiefly as snow. A new ice sheet would begin to grow in Canada. Therefore, the temperatures in Europe would fall and the West Antarctic Ice Shelf would melt, with the already mentioned consequences of increasing sea level. In order to prevent this climate catastrophe, a dam across the Straits of Gibraltar is proposed, effectively to steer the flow in and out of the Mediterranean. Experts easily disproved the hypothesis; yet one could read and hear about it amply in the media.

To this list of alleged climate catastrophes must be added, of course, global warming due to anthropogenic emissions of such greenhouse gases as carbon dioxide, methane or CFCs. As indicated, Svante Arrhenius described these mechanisms for the first time in 1897. Already long before the present discussion, there were suggestions that actual warming trends existed and were caused by the anthropogenic Greenhouse Effect. The American meteorologist Joseph Burton Kincer (1874–1954), past president of the American Meteorological Society, drew attention to unusual warming trends in 1933 in the *Monthly Weather Review*, and the British engineer and amateur meteorologist Guy Stewart Callendar (1898–1964) conjectured in 1938 in the *Quarterly Journal of the Royal Meteorological Society* that the trends then had to do with an elevated carbon dioxide concentration.[65] Shortly after the publication of this finding, temperatures began to fall, so that in the 1970s the American climatologist Stephen Schneider (1945–) spoke of the danger of an imminent large-scale cooling and even ice age.[66]

The much-discussed problem of the "Ozone Hole" is hardly a climate problem, strictly speaking. It concerns a change in the composition of the stratosphere, that is, the atmosphere above 10 km that influences the

---

[65] J. B. Kincer, Is our climate changing? A study of long-term temperature trends, *Mon. Wea. Rev.* 61 (1933) 251–259.

G. S. Callendar, The artificial production of carbon dioxide and its influence on temperature, *Q. J. Roy. Met. Soc.* 64 (1938) 223–239 also J. R. Fleming, The Callendar Effect (Boston: American Meteorological Society, 2007).

[66] S. I. Rasool and S. H. Schneider, Atmospheric Carbon Dioxide and Aerosols: Effects of large increases on global climate, *Science* 173 (1971) 138–141. Today, Schneider is one of the most influential environmental biologists and climate scientists, who vigorously warns about the consequences of global warming and demands political action.

filtration properties of the atmosphere. In the case of a hole in the ozone layer, less ozone is in the stratosphere; more ultraviolet radiation passes through the atmosphere to the surface of the earth and, as a result, harms the health of people, animals and possibly also plants. The ozone hole is also classified as an anthropogenic phenomenon, because the destruction of ozone leads back to the advance of exclusively artificially produced chlorofluorocarbons (CFCs).[67]

4.7. The Influence of Climate Changes on Society

A determination, or just a scenario, of the effects of climate change on the natural foundations of life, the economic, political and social consequences, and the possible feedback on climate, is an extremely complex problem.

In this section the problem is dealt with as follows: first we will discuss the anticipated direct consequences, such as agricultural productivity, coastal protection or the spread of tropical diseases. Then we will discuss the question of which options a society has to deal with such changes and what a rational climate policy might look like. This can occur in the framework of the "Global Environment and Society" (GES) model. Differing from this rational GES model, the "perceived environment and society" (PES) model no longer describes a rational, informed society, but rather a society that comes to decisions on the basis of socially constituted conception of climate change, including many uncertainties and their effects. While the GES model is mathematically formulated, the PES is a qualitative model.

Let us first deal with the consequences of climate change, with "climate change impact".

In this connection, the most important question concerns the time and location of expected climate changes. A frequently used characteristic measure of the anthropogenic climate changes is the *global* average air temperature. This quantity functions meaningfully as a global indicator of the intensity of climate changes, but it is practically irrelevant when an

---

[67] Cf. Reiner Grundmann, Transnational Environmental Policy. Reconstructing Ozone (Routledge, London, 2001).

understanding and assessment of regional impacts. Only these regional and local effects are of importance for society as well as for the ecosystem. In some areas the temperature will rise more quickly than in others; in a few areas it may even cool down. Overall, one expects an intensification of the water cycle of evaporation and precipitation, so that more rainfall seems possible globally. The precipitation distribution could be shifted, so that some areas will be moister in the future and others dryer.

Another often-discussed impact is on the sea level, which is expected to rise because of thermal expansion of seawater. Other influential variables on the sea level rise are the melting of glaciers and the decreased or increased rainfall on ice sheets. Regionally, the wind conditions, ocean currents and the changing gravity fields related to melting ice sheets will also influence the sea level. The media — and also, in particular, the insurance industry — often mention that much stronger winds are to be expected, seemingly everywhere. Scientific investigation to date does not confirm this assertion.

The assertions of global changes rest on calculations with climate models approximating reality, which can make estimations for areas of many hundreds of kilometers and more in diameter. But in climate impact research, mostly regional and local assessments are required, for areas with diameters of less than 100 km and under the consideration of, for example, politically relevant borders. The present climate models cannot provide such assessments, although the results of the models are often presented in the form of maps that, to lay people, also suggest local details to be read. Certainly local assessments may be drawn from the model output with the help of "downscaling" methods, utilizing the fact that the local climate may be understood as a joint effect of large-scale climate and local geographical detail, such as topography or land use.

The timing of states, i.e., the association of a model state with a specific time, is also problematic, because the climate model output represents a scenario, i.e., a plausible possible future state, but not a definite forecast for a specific time, like a weather forecast. Rather, the model scenarios are of the following kind: in case the emissions of greenhouse gases develop, as was supposed by the economists of the IPCC, then presumably a continuous temperature rise will emerge in the future, overlaid by natural climate variations unknown in their details. No one can say whether the summer of 2013 will be warmer or cooler than the average of the previous 10 years. Chances are that it may be warmer, but uncertainty remains.

A check of indirect and direct consequences of climate change mentioned in the media and in scientific literature indicates that the impacts of climate change regarded as possible appear omnipresent, similar to the discourse of climate determinism.

It is generally expected that climate change, say in terms of temperature or rainfall, will significantly affect agriculture. A shortened period of frost can lead to more frequent or intensified parasite attacks on flora and fauna. A few areas will experience worsened agricultural conditions, while the situation can improve in others. The current habitat borders of flora and fauna will be displaced poleward as a consequence of the global rise in temperature. To what extent the ecosystems in certain regions could adapt themselves to altered climate conditions also depends on the intensity and speed of climate change. Altered precipitation will influence the water supply and partially endanger it. One expects a fertilizing effect from raised atmospheric CO_2 for many plants, with the positive effect of increased agricultural production. How far future technical innovations, combined with new kinds of cultivation, will make possible new adaptation strategies is unclear. Because people will not react passively to a climate change, in the course of evolving climate change learning processes will certainly be set in motion that will mediate the effects considerably.

When we spoke to agricultural organizations in Germany, we observed little concern about climate change, at least for the foreseeable future. There were two reasons. The first is related to the fact that adaptation, by changing management concepts or by choosing different crops, which may also be bred to meet certain climatological and other conditions, can be carried out in a time frame of a few years, which is a short time scale compared to the speed of climate change. The second reason is that agriculture is influenced by other factors, which are considered more important, such as European agricultural policy or the demand for crop-based fuels. Of course, the situation may be different in other parts of the world.

Generally, a rise in the water level is expected to occur as a result of the Greenhouse Effect. Various studies say that with an average temperature increase of 2–6°C until the year 2100, sea level may rise a half-meter to one meter by 2100. This number accounts for the so-called thermal expansion of the sea water. If water warms up, then it expands slightly — which, becomes visible, however if we consider a large body of water, such as the ocean, and just look at the surface!

An open question is what may happen in the future with the large chunks of solid water on Earth, namely Antarctica and Greenland. The fate of these ice-blocks depends on two factors — namely how much is melted along the lower margins, and how much mass is deposited (by precipitation) on the higher levels. At this time, it is still a contested issue how the factors will steer the mass balance of the ice sheets. Satellite-based observations, which so far exist for merely a few years, are insufficient for a final conclusion. Thus, again, the jury is still out...

Even if the warming comes to an end in 2100, the sea level will continue to rise, because the warming of the seawater lags behind the atmospheric warming. An extreme scenario has been constructed for the Netherlands, according to which a rise in sea level, which includes the effect of a partial melting of Greenland and Antarctica, describes an increase by 1.3 m or so by 2100 and 2.5 m by 2200 as possible given our present limited knowledge, although not as probable.

What do such increases mean for coastal defense? A naïve approach, but one that is sometimes used by interested parties, is to assume that the sea level would simply flood the coastal plains as much as is geographically possible, irrespective of coastal defense. An example is shown in Fig. 25, for Northwest Germany along the southern coast of the North Sea. In this case, a sea level rise of 5 m is assumed, for two reasons. First, such numbers float in the public discussion, and second, it is the height of a heavy storm surge in the part of the coast. That is, such events happen every now and then, without flooding the coastal plains.

The diagram indicates that large chunks of the coast would be lost, including large cities like Bremen. But as the experience with heavy storm surges demonstrates, modern coastal defense is able to avoid the dire consequences of surges, at least in industrialized countries. When used as illustration for future stakes, then diagrams of the sort of Fig. 25 represent first of all disinformation for the (inland) public, who often has no clue of what the coastal dangers are. For instance, the Netherlands have dealt successfully for centuries with the rising water level as a result of sinking land.

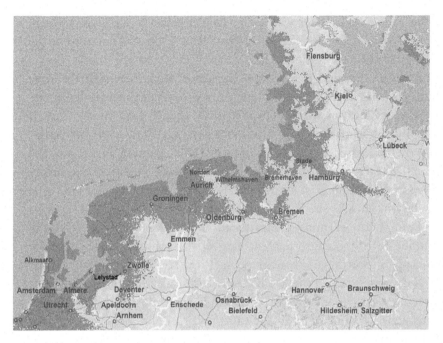

Fig. 25. What remains of the German and Dutch North Sea coast if mean sea level rises by 5 m... and all coastal defense is taken away.
Source: http://flood.firetree.net/

Sometimes threats to harbor and offshore installations and the like are also listed as possible detrimental climate change impacts. Certainly the typical lifetime of these sorts of installation amounts in any event to only several decades; so that an adaptation to changed water levels should be possible in the course of normal improvements and reconstructions.

Having said this, we add that sea level rise is a serious issue that needs our attention, but this attention should be professional, and directed towards improving the adaptation of coastal areas and populations to the dangers of the sea. An event like the disastrous storm surge in Myanmar in April 2008, which was associated with the well-predicted tropical cyclone Nargis, had nothing to do with global warming, but was a rare but regular event, which illustrated the poor conditions of coastal defense in many parts of the world. Sometimes people overlook the unacceptable lack of preparedness against present dangers, while drawing all attention to future dangers.

The anticipated climate changes could have a multitude of direct and indirect health consequences. A study by the World Health Organization (WHO) in 1996 warned of a significant rise in illnesses and contagious diseases as a result of the global climate change. The future weather, as the WHO deduces, favors the increase of bacteria, viruses and such disease vectors as insects and rats.

Considerable changes in rainfall — raised precipitation amounts in dry zones as well as droughts in previously moist areas — would lead to the spread of cholera, yellow fever and meningitis. Malaria could drastically increase because the warming of the earth by, for instance, 2.2°C would widen the distribution range of the Anopheles mosquitoes of today by 42% to 60% on the surface of the earth. Among the direct effects, so it is argued, there is also the rise in illnesses and cases of death caused by the increase in extreme weather regimes, such as intense heat waves. A study published in 1996 by the "World Watch Institute" expects an accelerated spread of epidemics due to environmental changes.

South American scientists warn that there is a risk that climate change may enlarge the geographical distribution of diseases like dengue fever, malaria, leishmaniasis or Chagas' disease, and that the breeding season of the vectors that transmit them may be prolonged.

However, not all scientists agree with this dramatic perspective. Paul Reiter,[68] the former chief entomologist at the US government's dengue research lab in Puerto Rico, points out:

> "Most people think of malaria as a tropical disease. That's completely wrong. Until very recently it was widespread in Europe and North America. In the 1880s, virtually all the US was malarious, and even parts of Canada. When... the Center for Disease Control and Prevention (CDC) was founded in 1946, its principal mission was to eradicate malaria from the US. In Europe, the disease was endemic as far north as Norway, Sweden and Finland. In the 1920s, epidemics killed hundreds of thousands in the Soviet Union, right up to the Arctic Circle. One of the last European countries to be freed of the disease was Holland. That was in 1970. As for dengue, the

---

[68] P. Reiter, Climate change and mosquito-borne disease, *Environmental Health Perspectives* 109 (2001) 141–161.

principal vector has been living happily in North America for about 300 years. At times the disease has been rampant. Indeed, the world's first recorded epidemic was in Philadelphia in 1780. In 1922, all the southern states were affected. There were an estimated 500,000 cases in Texas alone."

Reiter argues that climate would be only a minor factor limiting or favoring malaria. Instead, the organization of society and the precautionary health measures taken would be much more relevant. Indeed, the debate about the impact of climate change is also always a debate about the relative importance of environmental conditions which are, at least in the short term beyond the control of all of us *versus* the importance of social organization, for instance the organization of every day life subject to the control by at least some of us.

Socio-economic factors are far more significant than climate in determination of disease prevalence. For example, there are *one thousand times* more cases of dengue in the Northern regions of Mexico than in Southern Texas.[69] The climate across this 100 km band is the same, even the vector habitats are similar in many instances, but the pattern of social interactions and access to public health are vastly different. Socializing outside at dusk, when the mosquitoes quest for food, is prevalent in Mexico. North of the border, the people are indoor in air conditioned rooms, *socializing* around a tv set. Air conditioning is an adaptive measure, just like sitting outside in the cool evening air. One exposes the public to a health hazard, the other can protect them from vector-borne diseases as well has heat-waves.

Again, we suggest taking the issue of health effects very seriously. But this attention should be based on solid methodology, interdisciplinary perspectives and not naïve approaches such as attempts to determine the spread of malaria according to climatic indices. These simple geographical approaches share two facts, namely their appeal to naïve thinking and their disregard of other significant factors, chiefly from the social world. So far, the social sciences and humanities have failed to contribute enough to the

---

[69] D. Gubler *et al.*, Climate Variability and Change in the United States: Potential Impacts on Vector- and Rodent-Borne Diseases, *Environmental Health Perspectives* **109** (2001) 223–233.

scientific endeavor of disentangling the significance of climate change on social and ecosystems.

Impacts of changes in the natural environment and climatic foundations of human life can be stated only with difficulty, and with great uncertainty. Still more difficult is the assertion of societal, cultural and political consequences, not only in specifying future dangers, but also in the societal reactions to the predictions of such dangers — independent of their validity. Of course, we know that there will be impacts, but the difficulty is to determine what they might look like, and how we might mitigate the effects of such impacts.

Today, because we do not experience anthropogenic climate change in our everyday lives, but must rather rely on the complex analyses of the climate researcher, the social and political reactions rest exclusively on these future projections and their interpretations; our concern is not based on changes which have already occurred and are evident for everybody. Climate policy is not a reaction to climate change, but rather a reaction to the expectation of a climate change.

The prognosis of the impacts of climate change, of society's response to a climate policy, and of the effects of a climate policy on our economic system is consequently extremely difficult, if not impossible. One cannot simply project given social value systems, techniques, structures and processes into the future. This is particularly true in modern societies, which are not only characterized by a rapid, growing tempo of social change, but also by the growing incapacity of large institutions, like the government or science, to define and put into motion suitable, socially acceptable solutions for many occurring problems. Modern societies are knowledge societies, and social structures in which the influence of small social groups grows.[70] With the growth of influence of such diverse groups and social movements, it becomes more and more difficult to succeed in achieving social changes, for instance, an internationally agreed-upon climate policy. As far these ideas are concerned, it is to be feared that any adopted climate policy will be less consistent and reliable than the climate itself.

---

[70] N. Stehr, *The Fragility of Modern Societies: Knowledge and Risks in the Information Age* (Sage, London, 2001).

Among the most direct social consequences of climate change are above all the necessity for and shaping of climate management and climate policy. At the Conference of Environment and Develop of the United Nations in the summer of 1992 in Rio de Janeiro, the heads of nations and governments were convinced of the necessity to introduce measures to protect the climate. This agreement signaled the beginning of a global climate politics.

However, it turned out to be difficult in the following years to agree upon concrete steps, such as a specific limitation of greenhouse gas emissions or the accounting of emissions versus sinks, such as forests. The interests opposing measures to reduce emissions are very large, in particular in the world's worst emitting countries, namely the United States of America and, since recently, China. States whose economies are based mostly on oil or coal usage have no pressing interest in reducing emissions. On the other hand, societies whose economic performance is comparably weak expect advance concessions from the wealthy countries.

If we assume that we have a perfect knowledge base about climate change, climate sensitivity to changing conditions and climate impact, then the "technocratic" approach would be useful. This approach can be summarized in Hasselmann's model of the "Global Environment and Society" (GES; see Fig. 26).[71] According to this concept, human economic activity creates both valuable products and, at the same time, environmental stresses. These environmental stresses could be the emission of greenhouse gases, or heavy metals like lead or mercury, but also fire clearing of tropical rain forests. The environmental stresses act on the box "Environment" and cause changes there, such as degradation of soils or an elevated sea level. These environmental changes are felt again in the box "Economy, Society" and make countermeasures necessary there, such as changing agricultural practices or improving coastal protection measures. These countermeasures claim resources that could otherwise be used for the production of consumables and services.

---

[71] K. Hasselmann, How well can we predict the climate crisis? in H. Siebert (ed.). *Environmental Scarcity — The International Dimension*, JCB Mohr, Tübingen, pp. 165–183.

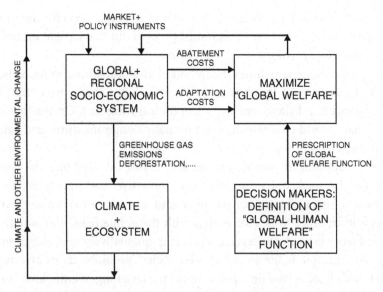

Fig. 26. Hasselmann's "Global Environment and Society" concept.

The effect of environmental damages diminishes economic performance as a consequence.

In this concept, there is the possibility that policy is steering the process to some extent. Tools available to that end are, for instance, environmental taxes and prohibitions, as well as orders. All such measures bind economic resources, but mitigate the environmental damage. The best political strategy is that which makes possible a maximum economic performance. This performance is measured by the sum of two costs, namely the expenditures to mitigate environmental damages ("abatement costs"), and for adapting to environmental damages ("adaptation costs"). Thus, the problem to be solved is an optimization problem, under the assumption of a worldwide consensus about climate change and climate policy.

The GES model can be formulated mathematically: to describe the climate components, strongly simplified models are derived from regular complex climate models. For the specification of the adaptation and abatement costs, simple assumptions about the functions of the overall economy are made. One of the GES models attained political significance when the White House in Washington used it in its preparation for the Rio de Janeiro conference in 1992. This model pointed out that a slight diminishing of the CO_2 emissions would be optimal.[72]

The GES statement assumes the fiction of an international consensus, or a world government that has complete knowledge about present and future processes and their sensitivity to political measures. On this basis the imaginary world government can rationally design measures and can also put them in motion.

The assumption of "world government" is not a crucial one and can be replaced by weaker assumptions. An alternative assumption is that there are a number of contract partners who all seek to maximize their own benefit. In order to be able to deal with this refinement, mathematical game theory is used. However, in this setup qualitatively new questions emerge. An example is the so-called "Free Rider" problem: let us assume that 100 states have agreed on a protocol for the lessening of emissions. All of the states have a part in the emissions and all suffer from the resulting damages. Through the protocol all states assume expenditures for reduction of the emissions, but also have benefits. If only one of the 100 states should decide not to abide by the protocol (thereby becoming a "Free Rider"), this would only affect slightly the whole emissions reduction; all states, including the non-abiding country, would continue to profit from the benefits of the protocol. But in our example only 99 of the 100 states would carry the costs for these benefits. The hundredth state has no direct costs, but rather only the benefits; it could perhaps achieve still further gains by raising its emissions. This argument is valid for each of the participating states, which has to ask whether it is really sensible to sign the protocol. In each case there are good reasons not to do so.

The need for "complete knowledge" of all relevant information can also be weakened in the model. One can postulate statistical uncertainties for a series of factors in the GES model, such as uncertainties with regard to future costs or presence of natural, slow climate changes. The mathematical problem then becomes a stochastic optimization problem. The statistical representation of inaccuracies requires certain assumptions,

---

[72] W. D. Nordhaus, To slow or not to slow: The economy of the Greenhouse Effect, *Econ. J.* **101** (1991) 920–937.

such as an assumption that on average an anticipated parameter, like costs for mitigation, or sensitivity, like the damage per degree temperature change, is valid. However, such an assumption is not consistent with social reality: one example is the possible growth of fundamentalist religious movements that put into practice the biblical commandment, "Subdue the earth and multiply". Such a development would make necessary a dramatic redefinition of the "cost functions" that can hardly be represented as a statistical variation.

Another problem with the GES model is that, ideally, complex and realistic climate models would be employed in conjunction with economic models of similar complexity and reality. But this is hardly possible. First, it is practically impossible to solve the optimization problem with complex models. Instead, simplified models must be used. Second, it is problematic, as already discussed, to deduce local assessments from the contemporary global climate models; but regional and local assessments are needed because climate damages are taking place regionally and locally. Finally, other than climate models, realistic economic models are, because of their empirical design, bound to contemporary structures and processes, and are valid only for a relatively short time horizon — only for a few decades, even in an optimistic case. Because the periods of climate change are many decades to centuries long, realistic economic models are not really usefully applicable in this context. Therefore, the participating models must be highly aggregated, i.e., the description of many processes and components is combined and simplified, so that such details as the spectrum of national economic branches and their interdependence can no longer be represented. Thus, the GES model is primarily a demonstration tool to explain the basic structures, but it fails in the concrete specification of necessaries and solutions.

Besides the methodological objections to the GES model advanced in the previous section, there is still a series of fundamental problems. The most important is that in the GES model, the social dynamics are represented exclusively in terms of economic concepts and economic value judgments.

A further deficiency of the technocratic approach consists of the supposition that our society and its value system react to "information" about the relevant, continuously developing "climate change signal", and correctly distinguishes this from "noise" produced by the ubiquitous natural climate variations and extreme events. This is certainly not the case. A demonstration of this difficulty of society to understand the issue of man-made climate is the observation that in the last year, all unusual weather regimes have been misused by the media and the public as proof of global climate change: the return of the "ice winter" to Germany in the winter of 1996/97, the meter-high snowfall in December 1996 that brought everyday life to a standstill in the Canadian cities of Victoria and Vancouver, or the floods and storms in the American states of California and Washington in January 1997, were commented on and interpreted by the media and by individual scientists as impressive evidence of global climate change.

Thus, on 4 October 1997 one newspaper reported:

"The record snowstorm of last week is a further reminder of the fact that we have experienced in the last few years the most variable and extreme weather ever observed. This is also true for other parts of the country and the world, such as the sequence of storms and snow melt in the Pacific Northwest in January and the record high flooding/inundations along the Ohio River last month prove. One can view this instability as early proof for the warming of the atmosphere and the oceans due to use of fossil fuels in the last 100 years."

In the winter of 1996/97 there were practically no storms in storm-accustomed North Germany; instead the winter weather was unusually uniform. The lack of winter rain led to dryness in spring, with increased outdoor fire danger — in this case no journalist thought of asking the climate researchers whether the failure of storms to appear could perhaps be an expression of the coming climate catastrophe. That the weather behaves crazily is normal. Whether the abnormal weather regimes lie outside the normal in their frequency or intensity can hardly be determined, because the natural variability of these events is enormously high.

It is also illusory to assume that with the help of dramatically simplified models of social dynamics, practical knowledge may be gained that makes it possible to design a climate dealing consistently with changes of living conditions on the regional, national or international level. The system's capability of learning, the inertia of institutions, the vested interests of a multitude of actors (for example, social movements, parties, international organizations and companies), moral and political conflicts, the ongoing loss of power of large social institutions, the unexpected dynamic evolutions of social, political and economic processes; all prevent models of this sort from leading to practically useful insights upon which concrete decisions may be taken.

The GES model could be replaced by the model of "Perceived Environment and Society" (PES), which is sketched in Fig. 27. It distinguishes itself from the GES model only through two further boxes: the "experts" who monitor and interpret climate variations and explain them to the public, and the "social interpretation" itself, the process within the public that adjusts the experts' explanations to the culturally constructed cognitive models of society.

In the framework of the PES model, the following situations — formulated in a simplified and short-hand manner — appear to be more likely:

1) Anthropogenic climate change develops slowly, and the public is prepared by trustworthy institutions: then the real, slowly evolving climate change signal will nevertheless not be perceived, or will hardly be perceived. But the public will convince itself of the reality of climate

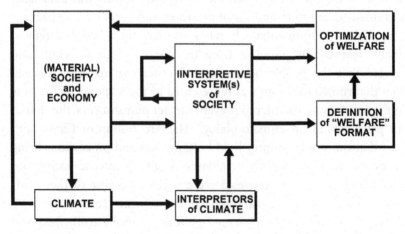

Fig. 27. The "Perceived Global Environment and Society" concept.

change by considering the naturally occurring extremes, which actually are unrelated to the systematic change, as evidence. An active policy of adaptation or abatement is conceivable; whether this policy will eventually solve the problems is another matter.
2) Anthropogenic climate change develops slowly, but the public expects no climate change: then a passive adaptation to the change takes place without consciously taking cognizance of it. The naturally occurring anomalous and extreme episodes will be correctly understood as natural events.
3) Climate is not changing due to human causes, but society nevertheless expects such a climate change: then, as soon as a natural event occurs consistent with the expectation, this is accepted as proof of the existence of ongoing climate change. Measures are taken according to the historically and culturally given beliefs.
4) Climate is not changing due to human causes, and society expects no climate change: then a "business as usual" policy will be adopted, independently of appearances of climate extremes.

Historically observed variant (4) is doubtlessly the most frequent case.

An example for case (3), "no change, but an existing expectation" is the English society in the years 1314–17. The harvests failed in these years, essentially due to incessant rainfall during the summer months. In the wake of this climatic anomaly famine developed, which was associated with dramatic death rates and social pressures and burdens. The particularly credible and influential churches of that time had beforehand repeatedly warned the populace from the pulpits of God's wrath, and demanded that the people behave in a God-fearing manner. The bad weather that ruined the crops was represented by the "experts" of the time as God's punishment. In order to prevent further punishments, the church put in place a kind of "climate policy". The Archbishop of Canterbury ordered that the whole country hold holy services and processions bring sacrifices, donate alms, and that everyone engage in intense fasting and praying. The result could be considered a success, because further summers were not spoiled by rain and the harvests returned to normal.

An example for case (2), a "change without advance warning", is given by the 1920s and 1930s, when the global average temperature within a few decades rose about 0.5°C. The change was described in a scientific journal in 1933, and in 1938 it was connected with the anthropogenic Greenhouse Effect.[73] But these findings and interpretations found no resonance in public discussion — undoubtedly because after the end of World War I, the worldwide economic crisis and the rise of a totalitarian system appeared much more important to the public than environmental changes.

The case of the presently ongoing global climate warming by greenhouse gases is an example of category (1). After initial discussions in scientific circles in the 1970s, in the 1980s the problem became the most important theme of climatic and environmental research. Extreme weather events, the growing environmental consciousness and multiple public warnings by scientists allowed the media and the public to take increasing interest in this problem. Parallel to this interest were growing fear and anxiety about the consequences of climate change.

Examples for such extreme weather events in the 1990s were:

According to reports in the US media, the drought in the U.S.A. in the summer of 1988 was causally linked to the anthropogenic Greenhouse Effect "with 99% certainty", as testified at a hearing of the American Senate by the American climate climatologist James Hansen (1941–). Hansen himself disputes this statement, whose indefensibility is demonstrated by the fact that after 1988 there were no further droughts of that sort. An alternative explanation attributes the drought in North America to a particular constellation of surface area temperature in the Pacific.

In Northern Europe, a series of storms in the winter and spring seasons of 1991 and 1993 attained prominence in the public debate about the reality of "Global Warming". Renowned scientists explained that the frequency or intensity of storms, for example in the North Sea, has systematically increased, probably due to the enhanced Greenhouse Effect. The assertion was later refuted, but the public hardly took notice of this refutation.

This list could easily be extended with examples in recent years, but it may be interesting for the reader to notice that the present practice of

---

[73] Kincer and Callendar, *op cit.*

blaming global warming for dramatic events has been fashionable since the late 1980s.

From what has been said so far we may conclude that society itself does not perceive climate and climate change as such, but rather heeds the warnings of the experts and their assertions, especially when related to extreme weather occurrences. Modern society perceives climate essentially through a socially constituted filter. This filtered image of the actual climate system we call the "social construct of climate". The interpretation of information about the climate is largely determined by the judgment of the experts from whom this information stems.

The reliability of the experts (in our examples, from the sciences or the churches), the absence of graver social problems, and the manner in which these questions are presented (or not presented) in the media of different countries have a larger influence on the common belief and understanding of climate than real climate experiences.

In other countries, there are other problems seen as more important. In China, for example, one is more interested in providing households with refrigerators than in the effect refrigerants have on climate; in Bangladesh the fulfillment of basic needs of the population (for example, the protection against dangers related to tropical storms and their associated storm surges) has a higher priority than a possible rise in the mean sea level. The task of climate and climate impact research thus cannot be only research into natural climate variations and anthropogenic climate changes. Rather, a multidisciplinary approach must be undertaken as well, to analyze and understand the complex problem of the perception of the climate and its changes.

5

Zeppelin Manifesto on Climate Protection

Man-made climate change has become a regular headline not only in the science sections of the media but also on the front pages among the key political issues, on the economy pages and lifestyle sections. However, this reporting and discussing in the public is predominantly one-sided, with the support of influential circles within climate research.

Up to now, it is almost exclusively measures to do with energy, transport, industry and housekeeping that have enacted under the heading of climate protection; such as measures to save energy and to increase efficiency, and the corresponding legislative frameworks.

The threat posed to the basic living conditions of society by climatic changes cannot be combated, as it has been up to now, only by protecting the climate from society, particularly given that many of these measures are of a symbolic nature. Additional effective efforts are required on the part of researchers, politicians and economic leaders in order to come to terms with the climatic dangers that already exist today, and which will intensify in the future, even in the face of a successful climate protection policy. This protection cannot wait to be put in place only after we have lived through catastrophes in the wake of weather extremes; rather, they must be realized in the form of precautionary measures. And these are in short supply here and now!

Sometimes such a proposal is countered with the declaration that extending the existing climate protection policy by means of an active precautionary climate policy is essentially identical with admitting that the existing policies have miscarried. This argument is obviously short-sighted and unfounded.

Concentrating climate policy on the reduction of greenhouse gases serves no purpose, if it leads at the same time to preventing taking precautions. Such a one-sided research perspective and climate protection policy will neither protect the climate from society in the coming decades, nor society from the climate.

We claim that the following issues need to be taken into account when designing global and national climate managements:

1) Climatic warming is not a fleeting, temporary or short-lived phenomenon. It is important to state this outright, because the impression is often given, intentionally or otherwise, that the climate can be changed in one direction or the other in a short span of time.

    Lowering emissions means, in the first place, only reducing the increase in their concentration. And, in fact, it would already be a triumph if we were presently to reduce the increase of these emissions. The long-term prevention of global warming, however, requires a quite extensive reduction of greenhouse gas emissions, i.e., lowering human emissions to almost zero. The length of time necessary for our elevated concentration of CO_2 to return even approximately to its original (here: pre-industrial) equilibrium amounts to somewhere between several decades and a few centuries.

    Why are these time spans relevant? On the one hand, they point up the prodigious efforts that are necessary world-wide in order effectively to halt climatic warming; on the other, these numbers are the point of departure for our further theses regarding how society will have to deal with the consequences of climatic warming.

2) Adaptation and prevention, i.e., reduction of emissions, are reasonable options that must be pursued in concert. As a rule, however, they are different options. Adaptation to the dangers posed by the climate will only incidentally reduce emissions; likewise, energy-saving and other reductive measures will only seldom be able to reduce the vulnerability

of our basic living conditions in face of the dangers posed by the climate. What both options have in common, however, is that they are promoted by means of technological innovations, but most particularly by means of social changes. A realistic assessment and a public discussion of the dangers of climate change are the first prerequisites for understanding the nature and the extent of the social changes required. A positive atmosphere, in which innovations are actively promoted and publicly acknowledged, is useful not only in the context of an active climate policy.

3) Reductive measures are in any case reasonable and necessary. The same is also true of adaptive measures, which continue to have a lasting effect when the reductive measures begin to work at a later point in time. The more effective the reduction, the more efficacious the adaptive measures — in the long term!

4) Let us proceed, in a thought experiment, from the premise that human beings on this planet could manage to meet the goal of reducing emissions by 80 percent in the space of one year. When, under these conditions, would the climate machine achieve a new "equilibrium"? The answer is: not for decades. In other words, the climatic change that is already underway cannot be prevented overnight, even by the greatest imaginable efforts in the realm of mitigation policy.

A climate policy that commits itself to the problem of mitigation while neglecting the urgent need for adaptation is an irresponsible climate policy, because it denies society's inevitably higher degree of vulnerability in the coming decades. The goal of such a policy — to protect climate from society, and thereby to protect society from itself — will bear fruit only in the distant future.

A representative example of the prevailing one-sidedness of the discussion of climate protection and efforts in this area is the often dispassionately employed term "*heat deaths*". As if people were almost inevitably and defenselessly victims of nature, and not victims of specific social circumstances; and indeed of social circumstances that irresponsibly put people at the mercy of extreme heat and its consequences, and do not preventively shield the segments of the population that are most severely affected. To speak of "heat deaths", as was done in the case of the hot European summer of 2003, protects only the municipalities, regions or countries that failed in their duty to take precautions. The very use of this term guarantees, so to speak, that the trends that are the actual cause of this phenomenon will be thoughtlessly repeated.

5) There are at least three important reasons why politicians, society and scientists must urgently think in terms not only of mitigation, but also of *precautionary* measures, as a reaction to the consequences of climate change:

The time scales of the long-term results of lowering emissions and of climate change do not correspond to each other. Any successes in terms of reducing the emission of greenhouse gases will take effect, as we have said, only in the far future. A world in which only small amounts of CO_2 are still being emitted will come too late to limit climate change in the next decades. The practically unlimited emissions of the past and up to now guarantee that climate change will change our future living conditions. The dilemma lies in the fact that the time scales of nature are not congruent with those of political decision-making cycles in democratic societies, which proceed in terms of election periods and cycles of attention, and which are reflected in the limited horizons of human action.

The threat posed by extreme climatic events, such as torrential rains, floods and heat waves, is already considerable today, and always has been in many regions of the world. One need only recall New Orleans in 2005; the storm surge of 1872 on the German Baltic coast or that of 1953 in Holland; or even Hurricane Mitch, which was turned to good use in the course of the 1992 negotiations in Rio de Janeiro. The vulnerability of our basic living conditions increases parallel to the growth of the global population in endangered regions, where growing segments of the population are marginalized without protection and, not least for reasons of political economy, become victims of extreme weather events.

6) The regions of the world whose basic living conditions will be particularly hard hit by the consequences of world-wide climatic changes are already demanding today, rightfully and increasingly vehemently, that the world must see to their protection, and not only to the protection of the climate.

World-wide climate policy is particularly clearly represented by the Kyoto Protocol. The Kyoto Process concerns itself almost exclusively with questions of reduction. The reduction targets of the Kyoto Protocol, which expires in 2012, will hardly be achieved. The successful execution of the Kyoto Protocol's so-called "Clean Development Mechanism" (CDM), in terms of the world-wide emission of CO_2, would by 2012 reduce the volume of world-wide cumulative emissions by about a week's worth, compared to the same development without Kyoto reductions.

For developing and emerging countries, particularly China and India, there is currently no obligation to reduce greenhouse gas emissions. We have no precise data regarding the greenhouse gas emissions produced by these countries, but we can assume that their share of the global balance of greenhouse gases is continually increasing. In the future, however, the developed societies will also emit (yet) more climate-damaging greenhouse gases. The total emission of carbon dioxide above all, despite all efforts at reduction, will probably increase further in industrialized countries between now and 2012.

The Kyoto approach, as a form of socially restrictive, large-scale global planning, has failed. Any subsequent process based on this hegemonic planning mentality will serve no purpose. As a result, climate change of human origin is steadily advancing, and will step up in the future. A reversal of this alteration to our global climate will be possible only over the span of decades, if not centuries.

7) Despite the contrary opinions of all political parties up to now and their reluctance to speak publicly about precautionary climate programs, adaptation as a precautionary measure is relatively easy to implement and to legitimize in political terms. Moreover, it has the enormous advantage that its success will be evident in the foreseeable future. When it comes to finding solutions to a problem by means of innovations in science and technology, it is easier to present these in the form of adaptive measures.

8) The consequences of warming vary significantly according to region and climatic zone. Research into precautionary measures thus means expanding our knowledge about regional changes. To what, exactly, are we going to have to adapt? With the aid of adaptive strategies several

goals at once can be achieved, because they are primarily locally or regionally oriented, and therefore can be flexibly configured: improving quality of life, decreasing social inequity and increasing political participation are not mutually exclusive.

9) The dual challenge of adaptation and prevention also leads to a reasonable division of labor. The national and European responsibility falls at the level of the frameworks for managing emissions, while for those in charge of the states, provinces, countries and municipalities, the question of reducing their vulnerability should have priority. In fact, institutions and persons charged with specific responsibilities — for coastal protection or for the Hamburg harbor, for instance — demonstrate a concrete commitment to solving problems of adaptation.

10) In the public discussion, down to the present day, prevention alone has been portrayed as a virtuous form of behavior, even when it takes the form of purely symbolic and largely ineffective actions, such as Sundays without driving, doing without long trips, or staging public events. This perception is not unproblematic, to the extent that it gives actors the impression that sufficient steps are being taken to protect the climate. A revision or extension of this perception to include a proactive attitude toward precautions and toward necessary social changes, however, as is essential to protect society from the changing climate and thus to reduce the vulnerability of the very basis of our existence, is still lacking. An effective defense of this basis demands precautionary measures in the coming years and decades. This must be our priority.

6

Summary and Prospects

One of the messages of this book is that the scientific subject matter "climate" should not only be located within the domain of natural sciences, but also within the realm of the social sciences and humanities. This is even more valid when the public and policy-makers have to be advised how to deal with suggestions and warnings prompted by scientific climate research.

Scientific climate research, as is the case for all of the natural sciences, is generated within a particular socio-historical context. Most recall Svante Arrhenius, only a few remember Guy Stewart Callendar, and Eduard Brückner is almost forgotten.

In the zeal of battle and in the excitement about new work, an event, a discovery and an idea of only a few decades past is quickly forgotten. If past scholars are fortunate their contributions may have been incorporated into contemporary scientific discourse by obliteration.[74] Mistakes and misconceptions are not remembered, with the danger that they will be repeated with the same enthusiasm as in the past.

---

[74] A concept introduced by Robert K. Merton in his *Social Theory and Social Structure* (1949). In the process of "obliteration by incorporation", both the original scholarly idea and the literal formulations of it are forgotten due to prolonged and widespread use, and enter even into everyday use, but no longer being linked to their originator; the idea or the discovery become similar to common knowledge in a field of discourse.

Social science research has effectively shied away from dealing with the results of the environment on society. After the painful experiences with climatic and biological determinism of previous periods, this is understandable. The consequence, however, is that there is hardly any social science climate research, and that the field of climate change is more or less exclusively covered by the natural sciences.

For centuries, the conviction that the climate is among the determining factors of human civilization and human behavior was almost an uncontested dogma. The doctrine of climatic determinism was a common component of scientific and popular explanation models. Today we are on very much more certain ground when we assert that the climate is not determining, but more or less conditioning, social processes.

At the same time, for its part, human behavior has an influence on climatic conditions. In historical times this influence was of a regional kind, such as through large surface changes of land use; today we are confronted with global changes. The past has shown that humankind will solve the problems linked to these changes, and we are convinced that it will continue to succeed in this. Mankind is solving its problem according to its own dynamics.

The expected global climate change should not be confused with "normal" extreme weather events. Because these extreme weather events are of great public interest, and are also linked often enough with serious consequences, they are often incorrectly interpreted by lay persons or the media as an unmistakable signal of climate change. A few journalists, scientists and politicians use these mistaken interpretations to further their own interests, which are not really related to the problem of climate change.

Human existence in the 20th and 21st centuries has been influenced by science and technology as in no other historical period so far. The dependency on science and technical artifacts grows constantly. Whether this is a blessing or a curse depends on the view of the observer. The once uncontested admiration, satisfaction and confidence in modern science and technology have waned and been replaced by a much more skeptical attitude toward novel scientific knowledge and technical capacities.[75]

---

[75] See N. Stehr, *The Governance of Science* (Transaction Books, New Jersey, 2004), and N. Stehr, *Knowledge Politics. Governing and Consequences of Science and Technology* (Paradigm Publishers, Colorado, 2006).

The 21st century will also see large changes in social values, processes and structures and in technology. Any reasonable climate policy has to anticipate these changes and their multilayered dynamics. Therefore, doubts about the chances of an effective, negotiated climate policy, especially in the sense of a broad-based, global mitigation of greenhouse gas emissions, are in order. But on the other hand, one can also confidently hope for a scientific-technological mastering of the climate problem through, for instance, new filtering techniques, alternative forms of energy and energy use, or improvements in energy efficiency.

How should we deal now with all of this knowledge? For scientists it should be a challenge to put into practice the interdisciplinarity of climate research. We need a "social natural science" that includes society as part of the earth's ecosystem, without reducing the non-mathematizable, internal social dynamic to an environmental determinism.

And as the Danish Queen Margrethe II suggested in her New Year's speech in 1998:

> "When looking at the many actual problems of the day, they often appear as confusing and much too complex: There is the problem of the uneven distribution of the resources of this world, the protection of the environment from pollution and ruthless exploitation of natural resources, with CO_2 emissions and the holes in the ozone layer. The situation is not made easier when we have to observe our experts having difficulties obtaining an overview, not to mention reaching unanimity about the problem. We must not let ourselves be swept along by sensation-mongering prophets of the end of the world; we should not allow ourselves to be driven like startled animals from one side of the corral to the other as soon as warnings about the wolf are called. That would be just as irresponsible with respect to future generations as pure onlooking. We must not refuse to deal with the great challenges of our time."

Index

Adaptation 29, 42, 53, 54, 114, 116, 121, 126, 130, 131, 133, 134
Acclimatization 44
Acid rain 98, 100
Albedo 33, 41, 82, 84
Anthropogenic climate change 3, 34, 64, 66, 70, 82–95, 106, 107, 110, 112, 119, 125–126, 128
Anthropogenic emissions (see also Greenhouse Effect) 86, 111
Aristotle 46
Arrhenius, Svante 32, 33, 111, 135
Astrology/astronomy 46

Beck, R. A. 58
Bernhardt, Karl-Heinz 55
Bjerknes, Vilhelm 35, 36
Brückner, Eduard 61, 62, 65–70, 96, 107, 135
Butterfly Effect 41

Callendar, Guy Stewart 111, 135
Chamberlin, Thomas Chalm 40
Chaos Theory 75

Climate catastrophe (see also weather extremes) 1, 66, 100, 102, 106–112, 124
Climate determinism 13, 50, 54–58, 61, 114
Climate models 8, 9, 42, 77–79, 85, 87, 93, 94, 113, 121, 123
Climate policy 3, 45, 112, 119–121, 126, 130, 131, 133, 134, 137
Climatic energy 51, 54

Deforestation 66, 69, 84, 85, 99–101, 104, 108, 109

El Niño 30, 74, 78
Emission scenarios 89
ENMOD, Environmental Modification Technique 79, 80
Eugenics 51, 53

Gaia Hypothesis 85
Galton, Francis 53
GES (Global Environment and Society) model 112, 120–123, 125

Global warming 1, 24, 54, 98, 99, 104, 106, 110, 111, 116, 127, 128, 130
Greenhouse Effect (see also anthropogenic emissions) 1, 8, 66, 83–86, 94, 98, 99, 105, 109–111, 114, 127
Greenhouse Theory 31, 33
Groven, Rolv 36
Gulf Stream 39, 78, 110

Hadley, George 34, 35, 38
Hadley cell 38, 42
Hann, Julius von 7, 13, 66, 68–70, 72
Hansen, James 127
Harmonic analysis 72
Hasselmann, Klaus 75, 76, 102, 120, 121
Health 5, 28, 47, 50, 67, 99, 112, 117, 118
Hegel, Georg Wilhelm Friedrich 47
Hellpach, Willy 57
Helmholtz, Hermann von 31
Herder, Hohann Gottfried 48, 85
Hildebrandson, H. H. 30
Hippocrates 46, 47
Hough, Franklin Benjamin 70
Humboldt, Alexander von 6
Huntington, Ellsworth 49, 50–55, 72
Hurricane Katrina 23

IPCC, Intergovernmental Panel on Climate Change 23, 87–89, 94, 95, 113

Kant, Immanuel 34
Kempton, Willett 97, 108
Kepler, Johannes 46

Kincer, Joseph Burton 111
Kyoto Protocol 133

Labitzke, Karin 73
Lauer, Wilhelm 56
Little Ice Age 74
Loon, Harry van 73
Lorenz, Edward 75
Lovelock, James 85

Margrethe II, Queen of Denmark 137
Milanković, Milutin 73, 75
Montesquieu, Charles de 47, 48

Neumann, John von 35
Nordhaus, William 54
Nuclear winter 110

Ogilvie, Astrid E. J. 44
Ozone depletion 100, 101

Pálsson, Gísli 44
Penck, Albrecht 65
Periodicities 72
PES (Perceived Environment and Society) model 112, 125
Pielke, Roger A. 23, 24
Plato 46

Race/ethnicity 47, 48, 50, 51, 53, 56, 109
Rebetez, Martine 96
Reiter, Paul 117, 118
Rossby, Carl Gustav 35

Saabye, Hans Egede 30
Schneider, Stephen 111

Scientification of climate 8, 18, 65
Sea level 87–89, 111, 113–116, 120, 128
Shaw, Sir Napier 72
Slutsky, Eugene 72, 73
Sombart, Werner 57
Sorokin, Pitirim 54
SRES, Special Report on Emission Scenarios 89–92
Stefan–Boltzmann law 82
Solar activity 69, 75, 84

Umlauft, Friedrich 49, 58

Virchow, Rudolf 43–44
Voltaire 45

Weather 2, 3, 7, 8, 11–14, 20, 22, 24, 26, 30, 31, 35, 39, 43, 45, 46, 51, 58, 67, 73, 78–82, 95, 96, 98, 100, 101, 106–108, 111, 113, 117, 124, 126, 127–129, 132, 136
Weather extremes (see also climate catastrophe) 12, 45, 129
Weather observations 46
WHO, World Health Organization 117
Williamson, Hugh 106

Younger Dryas 74, 75

Zeppelin Manifesto 3, 129–134